CULTURAL INFLUENCES IN ENGINEERING PROJECTS

CULTURAL INFLUENCES IN ENGINEERING PROJECTS

MORGAN E. HENRIE

 MOMENTUM PRESS

MOMENTUM PRESS, LLC, NEW YORK

Cultural Influences in Engineering Projects
Copyright © Momentum Press®, LLC, 2015.

First published by Momentum Press®, LLC
222 East 46th Street, New York, NY 10017
www.momentumpress.net

ISBN-13: 978-1-60650-482-6 (print)
ISBN-13: 978-1-60650-483-3 (e-book)

Momentum Press Industrial, Systems, and Innovation Engineering Collection

DOI: 10.5643/9781606504833

Cover and interior design by Exeter Premedia Services Private Ltd., Chennai, India

10 9 8 7 6 5 4 3 2 1

Printed in the United States of America

ABSTRACT

Culture can be a significant contributor to, or hindrance to, a team's success. Research has clearly established that failing to have a cohesive team culture creates a severe challenge to any team effort. Culture is also something that everyone brings with them to the team. Yet, developing an understanding of what the team culture is, what constitutes a cohesive team culture, and how to modify it such that it enhances the probability of team success is a challenge to team leaders. Cultural team challenges exist within holistic, that is, teams from a single nation, or multinational teams.

Cultural Influences in Engineering Projects provides team leaders and interested individuals a cohesive source of information, ideas, and approaches on how to understand, analyze, develop cultural transition plans, and methods which can improve or modify a team's culture toward success. *Cultural Influences in Engineering Projects* also includes an extensive literature review reference set which provides the reader a ready source where they can continue to expand their cultural knowledge base and ultimately improve their probability of successfully managing holistic and multinational teams.

KEY WORDS

artifacts, assumptions, behaviors, beliefs, collectivism, communication, convergence, country, cross-cultural, cross-culture, crossvergence, cultural, culture, culture transformation, divergence, engineering, ethics, group, heroes, hybridization, individualism, multicultural, multinational, nation, nation-state, norms, organizational culture, power distance, projects, scientific management, state, sub-culture, systems science, team, technology, traits, uncertainty, values

CONTENTS

LIST OF EXHIBITS

FOREWORD

I first worked with Morgan Henrie when he was an adjunct professor at the University of Alaska Anchorage and I was chair of the Logistics Department there. We traveled to Korea, China, and mostly the Russian Far East. We became research partners on a major project to develop and deliver project management curriculum to Russian oil engineers and later to design and build three logistics departments and programs for Russian universities in Vladivostok, Khabarovsk, and Sakhalin. It is during these five years of projects that we learned valuable project management and team behavior cultural lessons. What we discovered was totally surprising and unexpected. The Russians had no clear concept of the two terms—project and management—being used together as Morgan offers in this latest book. When we would take the Russian engineers to visit a pristine environment and pipeline facility in Alaska, they at first thought this was a show place, kind of like a Disneyland, just to show tourists.

Morgan has a wide range of experience with successful projects in the oil and gas and construction arena, plus extensive field research in how to manage different project teams scattered across the United States and overseas. What Morgan has packaged in this book is more than facts, definitions, and teachable explanations of project and program management. He has woven his experience into each segment, each element of how to see and react to managing a team of individuals, how to work and manage an organization as a project is developed and properly closed out at completion. If you are a novice to the field of program management or project management, you will readily grasp the basics with his examples for our everyday life, building a house or back deck or writing a book. Yes, even this book followed the basic principles of project and team management. If you are a professional engineer or project manager, then you will find the key performance indicators needed to expand your business overseas or work with a more diverse team or organization. If you are a teacher, this book is organized to help develop a useful curriculum from vocational training to a BS or MS degree. For the professionals seeking certification credentials, this book is a must-read; it will put your mind on the complex issues often overlooked.

The funniest part of Morgan's history in writing this book has to be during one of his five-week project management education classes for his first group of Russian engineers. Morgan was part of a contract to bring Western-style project management to these Russians. Twenty-five seasoned engineers were listening intently for several hours that first morning as Morgan described project management from the Pyramids to the Twin Towers to the Alaska oil pipeline. About two hours into the lecture, he called for questions. One senior Russian raised his hand and asked, "So, what is project management?" Stunned, Morgan queried the gentleman about projects and management. Later, he raised his textbook in the air and asked all the Russians to do the same. Then he tossed his book across the room to the trash can and asked the Russians to do the same. You see, Western textbooks had it wrong. They did not understand the different cultural impacts in constructing a building, running a team of workers, planning for risk, managing a budget, and so forth. Morgan picked up a marker and started writing on the white board. He had begun the first culture-based project management class. For a Western-trained project manager to visit a foreign country, not just the Russian Far East, you really need to know the national culture as well as the individual and organizational culture. Otherwise, your great plans, and wonderful college degrees and experience in the United States will lead to unexpected failure. Your success as a program manager, project manager, or team manager depends on really taking this book to heart. It will save your company, save your reputation.

I wrote this foreword for one simple reason: I trust and respect Morgan Henrie in his project and program management skills. We co-taught executive courses as well as undergraduate and graduate courses at the University of Alaska as well as established three programs in the logistics side of project management in Russia. His PhD dissertation from Old Dominion University laid the foundation for this body of knowledge. Whether it is the culture of workers from Virginia and Alaska or a team of European, Asian, or Latin American engineers, you owe your company and yourself a favor by reading this book; it can save your project.

Dr. Oliver Hedgepeth
Program Director and Professor,
Government Contracting and Acquisition American Public University
Author, *RFID Metrics: Decision Making Tools for Today's Supply Chain*, CRC Press, 2007.

ACKNOWLEDGMENTS

Creation of a book may seem like a solitary activity where the author appears to be alone in his endeavor to bring forth the final product. Yet, there are many people working side-by-side with the author to create the product the reader sees. Most of those who support, encourage, and assist along the way generally remain anonymous. While the generation of this work required the assistance and support of more people than I have space to acknowledge, I want to specifically identify and thank Dr. William R. Peterson for inviting me to participate in this series of books. His invite started all this moving forward and his input and support along the way kept it moving.

I need to especially thank two key reviewers whose insight, suggestions, and guidance enhanced the final outcome. First there is Dr. Oliver Hedgepeth who is an American Public University Professor and Program Director of Government Contracting and Acquisitions program. Then there is Dr. Gamze Karayaz an assistant professor at Isik University Economics and Administration Sciences program. Even though both of these individuals have many demands on their time, they advanced this work by taking the time and exerting the effort to provide essential and critical reviews and comments.

Finally, but not the least, there is my family and publishing team who made sure I kept on track and didn't lose steam in obtaining the end goal. They know who they are and I remain eternally grateful for their support, encouragement, and assistance.

CHAPTER 1

INTRODUCTION TO PROJECTS, PROJECT MANAGEMENT, AND PROJECT MANAGERS

"This feeling, finally, that we may change things – this is at the centre of everything we are. Lose that ... lose everything"
—Sir David Hare (1947)

1.1 INTRODUCTION: PROJECTS—WHAT ARE THEY AND HOW ARE THEY DEFINED?

In this chapter, we lay the foundation for discussing cultural influences and effects within a technical team environment using the project as the team environment focus. To accomplish this, we first discuss what a project is, what project management is, and then project managers' responsibilities. This foundation is essential to understanding the broader context of this book: The team's cultural influences on their project.

To establish an early reference point on what we mean by the word culture, we view it as a social product which originates from social relations. The social lessons learned as we interrelate with those around us form the basis, either at the conscious or subconscious level, of how the society, organizational members, company employees, and project team members successfully interact with each other. Expanding on this further, as will be discussed in more detail later, there are many definitions in use on what culture means and no single definition has universally been adopted within the management discipline. While the management discipline has not embraced a single definition, we set the stage for a common understanding throughout this book by adopting the following;

"Organizational culture is a system of shared meaning held by its members that distinguishes their organization from other organizations" (Brady and Haley 2013, 40). Yet this shared meaning, as Schein presents, is an abstraction that "… points us to phenomena that are below the surface, that are powerful in their impact but invisible and to a considerable degree unconscious …" (Schein 2004, 8).

To understand the impacts that each individual's culture has on the project team, and therefore the project, one must first understand what projects are, the context you will find them in, the process by which projects are processed, their various organizational structures, as well as the project manager's role and responsibilities. That is, we must first lay the foundation of what projects, project teams, and project managers are before we can really begin the discussion on cultural influences within this environment.

Establishing the project context foundation refers to discussing and defining the various aspects of projects. It is imperative that one understand specific project and project management terminology, the various team structures that one will encounter, as well as what they are not. It is also good for the reader to obtain an introductory understanding of project management professional organizations and how they contribute to the advancement of project management as a discipline.

To set the stage of building a deeper understanding of projects, project teams, and project management we start by first looking at projects from a historical point of view.

1.1.1 PROJECTS—A SHORT HISTORICAL VIEW

In this section, we briefly trace projects from early recorded history through modern time. This walk through time allows us to see how projects have existed since the beginning of recorded time, and probably before, and how they have changed over the last 50 years.

In the beginning there have always been projects. Unique things had to be built, developed, or created. Someone had to come up with the idea and either implement or develop that idea or have others implement it for them. These unique events carried significant risk that they would fail. Carayannis, Kwak, and Anabari expand and detail the position that "Project management has been practiced for thousands of years since the Egyptian era …." (2003, 1). Validation of the idea that projects have always occurred is found across the globe and all we have to do is look for the historical evidence. When we look into Egypt, we see evidence of projects in the form of the pyramids. The Pyramid of Djoser is estimated to be the oldest of the known pyramids and dates from around 2630 **Before Current Era (BCE).** Building of these grand structures involved establishing the overall plan, developing detailed engineering plans, identification of proper building material, building site selection, obtaining the hundreds to thousands of labor resources, managing the construction material, food, and other resource supply chains, and someone making decisions on a constant basis. This constitutes a major project be it in 2630 BCE or in 2013.

Moving forward a couple of thousand years to 122 **Current Era (CE)** the Romans embarked on many significant project efforts including a major construction project called Hadrian's Wall. After about six years of construction effort, Hadrian's Wall stretched the width of England (73.5 miles). Scholars have suggested various purposes for the wall such as a fortification to protect the Northern Roman stretches from attack from those that lived north of the wall as well as a means to control immigration and exhort taxes from those passing through the wall.

Regardless of the actual end purpose or purposes, the construction of Hadrian's Wall was a major project. This effort, as with the pyramids centuries before, involved developing an architectural design; performing detailed engineering design; surveying the route; identification of raw material sources; managing the raw material supply chain; and arranging for, managing, and coordinating living conditions, food, and medical care for hundreds of construction workers for over six years, and someone had to be the ultimate manager who coordinated all of this and resolved the conflicts which naturally occurred. The managers also had to deal with ending the project and turning it over to the military for its final use. That is, they had to deal with management of change and termination of the project team.

If we move forward a little over 1,800 years, we come to 1931 and the completion of the Empire State Building, which for the next 40 years, was the world's tallest building standing at 1,454 feet with a total of 102 floors. The building supported thousands of people on a daily basis by keeping them comfortably warm or cool, in offices with sufficient light for them to work, and out of the blazing sun, rain, snow, sleet, and a host of other environmental extremes.

The Empire State building construction began on January 22, 1930, and continued endlessly until the grand opening ribbon was cut on May 1, 1931, just over 15 months in total. The construction effort involved 3,400 workers performing their duties in all weather conditions and cost, $24.7 million, in 1930, or about $372.9 million, in 2012. As with our other examples, this major construction project involved all stages of a project from conceptual idea to grand opening ceremonies as well as transferring the building from a project environment to a daily operations environment and dismantling of the project team. Building the Empire State building required someone to manage the overall effort, others to manage supply chains, quality control testing, and documenting and managing financial obligations.

As this short tour through history demonstrates, projects, as more fully defined in the later sections, have occurred for centuries. Yet, it appears that formalization of the project management process is a 20th century event. Prior to the 20th century, there is a lack of literature which discusses or details how these early projects applied the rigor, tools, and techniques associated with planning, tracking, and adjusting the project to meet specific cost, schedule, quality, or resource staffing as today's projects enjoy. As the next sections briefly discuss, these capabilities did not start to show up in the management area until Frederick W. Taylor, Henry L. Gantt, and others brought scientific management to the management field and the development of new and improved tools and techniques occurred.

Prior to Frederick W. Taylor's research, subsequent publications, training, and presentations, organizational management, as well as project management, lacked a definitive and scientific-based structure. The pre-Taylor era projects can be viewed as processes being performed based on a historical basis, this is how we do it, or based on individual assessments of how it should be done. There was no science or formal structure behind how work was scheduled or implemented and there were minimal, if any, metrics on how effectively and efficiently the work was being performed. The literature shows that this approach resulted in project cost overruns, late project deliveries, and projects which did not meet the user's needs.

Formalization of the management process, and subsequently the project management process, based on science and development of measurable metrics is linked to Frederick W. Taylor who "… is considered the author of *The Principles of Scientific Management* …." (Wrege and Stotka 1978, 736). Scientific management, as the name implies, is "… a scientific approach to managerial decision making… [which is] based on proven fact (e.g., research and experimentation) rather than on tradition, rule of thumb, guesswork, precedent, personal opinion, and hearsay" (Locke 1982, 14). Restating this from the historical view, prior to the development of the scientific management approach projects were implemented based on *this is how we have always done it*, a someone's personal belief in a better idea on how it should be done, and authoritarian decrees.

Changing from flying by the seat of your pants to applying scientific techniques and rigor which facilitates developing an understanding of the work process which provides one the ability to understand how the work is currently implemented, how it can be improved, and how to use quantifiable metrics to measure the change and continuously improve. This continuous improvement is based on the scientific principal of first developing an understanding of how things are currently being performed and how to quantifiably measure the current process with defined metrics. From this foundational understanding you can then develop how a new approach or method could be structured, planned, staffed, and implemented to achieve a higher level of effectiveness and efficiency. The new process is then implemented and the cycle starts over. Looking back to Taylor's early years, a classic example of this process is the study of how a company was loading coal. The study identified that much time was wasted in the current process, from the tools being used, to how work was assigned. The scientific management approach resulted in a major work flow change which included things like development and deployment of specific size shovels for specific sizes and types of coal and how workers were assigned jobs. These changes resulted in increased productivity of loading coal and improved processes (Istvan 1992). While loading coal is not a project, this example demonstrates the organizational process changes which scientific management generated.

During this same era, Henry L. Gantt, "… a close associate of Frederick W. Taylor …" (Wilson 2003, 430), was involved in the study of World War I (WW I) navy ship building. As a major construction activity and in alignment with other major production activities, Taylor and Gantt recognized that "The key to improving overall productivity lay in developing comprehensive planning systems" (433). Prior to Gantt's efforts and subsequent development of his planning system, ship construction occurred based on peoples' best efforts, historical information on what worked before, and when people and material were available. While the approach yielded completed ships, the resulting construction time and costs were longer and more than they needed to be. Filling WW I ship needs and cost constraints required a new approach which set the stage for Gantt's analysis and development of a new method. The resulting approach provided the various ship-building departments the means to coordinate their work to minimize resource conflicts (433) which improved overall efficiency and construction effectiveness. Ships were built in less time and at less cost. Gantt's approach bears his name today, the Gantt chart.

The Gantt chart, as we see it in today's form, is generally shown as a horizontal bar chart which identifies the start and end dates for major

Exhibit 1.1. Gantt chart example.

Activity	Monday	Tuesday	Wednesday	Thursday	Friday
Fabricate case					
Fabricate circuit board					
Fabricate display unit					
Assemble system					
Test					

activities; see Exhibit 1.1. Exhibit 1.1 is a very simple Gantt chart which shows the major activities involved in building a piece of electronic equipment.

From Exhibit 1.1, a person can see when specific activities are to start and when they are to end. While not explicitly shown or stated, one can also infer that a relationship exists between the activities. As an example, the system test will not start until after the Assemble System effort is completed. This relationship is inferred as testing is shown to occur on Friday and while the Assemble System activity is shown to end on Thursday. This implicit relationship provides input into the planning process. That is, a knowledgeable person or group responsible for testing can see and understand the overall process. This also provides them the information they need to coordinate with the person or group responsible for assembling the system. This radical approach provided the overall project team updated information which allowed for improved planning and ultimately shorter construction times and lower construction costs.

While the Gantt chart was first envisioned and used as a manufacturing planning and process control tool, it has evolved into many other uses to include project management. In fact, it is interesting that within today's project environment, project teams often refer to the project Gantt chart as synonymous with the overall project plan (Maylor 2001, 92). While development of a project's Gantt chart is not the overall project plan, it demonstrates the impact that the Gantt chart has had on projects' planning, monitoring, and controlling processes.

While the Gantt charts continue to be used today, the project management scientific approach continues to evolve. One major evolutionary project scheduling change was development of the Program Evaluation and Review Technique (PERT) and Critical Path Method (CPM) planning techniques.

PERT is referred to as a network planning tool with origins in the U.S. Navy Polaris missile program. Specifically, "PERT charts first came

into existence in 1958 by a company called Booz Allen Hamilton, Inc.....
Because ... the U.S. Navy wanted a simple system to manage and coordinate large and complex projects" (Stallsworth 2009). Prior to 1958 and development of the PERT and critical chain techniques, major government projects routinely experienced major delays, cost overruns, and failure to meet project objectives (Sinnette 2004). These negative outcomes were directly attributed to the project team's inability to visualize and document the detailed sequence of major project events and their relationship to each other. As such, as the project was implemented conflicts occurred with when and who was to do the work, when critical inspections had to happen, as well as when material was to be delivered. These conflicts resulted in late project deliveries, cost overruns, and failure to meet the quality requirements. PERT was developed as an enhanced planning tool which provided the project team the ability to plan and visualize the overall project activity and task sequence in the form of a network diagram.

A network diagram has many visual forms yet all share the common features of (a) providing a visual representation of the full project schedule, (b) showing start and end dates for all identified activities and tasks, (c) showing linkages between activities and tasks, and (d) providing a presentation of the project's critical path. PERT networks are different from Gantt charts in that PERT charts explicitly show the linkages between activities and tasks while the typical Gantt chart implies the relationships. Exhibit 1.2 is an example of what a PERT network chart can look like.

As Exhibit 1.2 shows, PERT charts provide significantly more details on the various project activities' relationships and durations than the project view demonstrated in Exhibit 1.1, the Gantt chart. It clearly shows the overall project and all the relationships between the various activities. It also provides the project team the ability to track the project progress as it moves forward in time by showing what work efforts have been completed to date as well what efforts should be starting and which ones are on the critical path. Development of a PERT chart requires detailed information which is obtained from the project work breakdown structure which is discussed later in this chapter.

Another significant historical breakthrough in project management planning, monitoring, and control was introduction of the critical path concept. To understand what a critical path is one must understand the overall project planning process. The overall project planning process is a set of actions which culminate in the development of a set of documents which detail how the project will be implemented. This is the project plan.

Development of the project plan begins by first identifying what the project objective is. As an example, if the project objective is to build a

ID		Task Name	Duration
	0		
1		Implementation Project - Tower Site	129d
2		Contract	4d
3		Contract Award	1d
4		Contract Review and Approval	2d
5		PO Issued	1d
6		Contract Design Review	7d
7		Pre-construction Conference	1d
8		A&E Site Design Review	1d
9		Tower Design Review	1d
10		Shelters-Comm & Gen - Design Review	2d
11		Design Approval A&E Drawing Changes	0d
12		Design Approval Tower Changes	0d
13		Design Approval Shelter- Comm & Gen - Changes	0d
14		Order Processing	5d
15		Order Tower	1d
16		Order Shelters	1d
17		Order Propane Tanks	1d
18		Manufacturing	47d

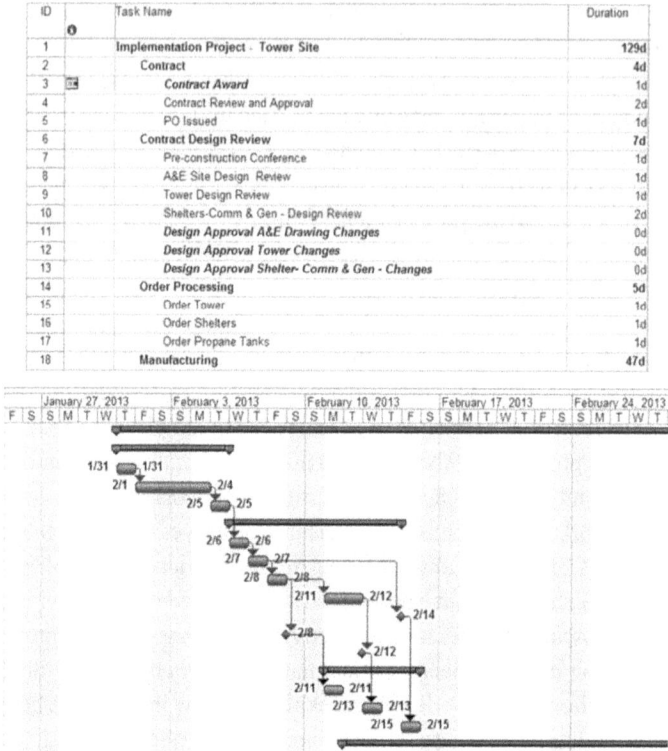

Exhibit 1.2. PERT chart example.

new sky scraper in downtown Houston, Texas, this has to be fully articulated and defined as part of the project plan objective or mission statement. Next, the project plan needs to identify specific and measurable objectives which will be used to evaluate the project during its implementation. An objective could be to complete fabrication of the building foundation within two months from start of construction or completion of the electrical rough-in and subsequent city inspections by a specific date. The keys to project objectives are that they are (a) specific, (b) measurable, (c) realistic, (d) include a time limit, and (e) agreed to by the project team, project owner, and all other key stakeholders.

Once the plan has been defined by the final outcome and measurable objectives, the plan needs to identify those things which are outside the scope of this effort and how scope changes will be managed. Defining what is inside and outside of the project's scope is essential if a successful project is to occur.

The project plan should also include a work breakdown structure. A work breakdown structure is defined as "... a set of related activities ...

with the assumptions that this set is necessary and sufficient to reach the project's desired results" (Levardy and Browning 2009, 600). A work breakdown structure is a hierarchical breakdown of the project into distinct activities and tasks where lower level activities and task completion are required to support the higher level activities and tasks. Exhibit 1.3 provides a general view of one work breakdown structure approach.

While Exhibit 1.3 is just a small part of this project's scope of work (SOW), it demonstrates that there are many tasks which must occur to complete the final objective of fabricating and installing a microwave tower. For each of the identified tasks, there are sets of subtasks and activities which must be completed before the higher level task is complete. (Note: Some project management literature reverses the use and definition of what an activity and task is. Rely upon the local project environment and culture to establish how you use these terms.) The work breakdown structure forms the foundation for a major project plan document, the PERT schedule and its inclusive critical path.

Development of the PERT project schedule requires that the project team add three primary data points to the work breakdown structure. The first data point is work effort duration; specifically, how long will each of the activities take to complete. The second set of data points is work activity precedence or linkages. This means that the project team must clearly define which activities must be complete before a different one can start, which ones can be accomplished in parallel, and so forth. As an example, in Exhibit 1.3 the tower cannot be ordered until the request for quote (RFQ) responses have been received, reviewed, and a vendor selected. The RFQ cannot be issued until all design and engineering work activities

1.0 Install a microwave tower

    1.1 Major Project Subsystem 1 – Procure all permits

        1.1.1 Task 1 – city permit

            1.1.1.1 Subtask 1 – submit form

                1.1.1.1.1 Work Package 1 – fill in forms

                    1.1.1.1.1.1 Activity - obtain city forms

        1.1.2 Task 2 – procure tower

            1.1.2.1 Subtask 2 – Analyze Request for Quote (RFQ) and select vendor

                1.1.2.1.1   Work package 2 Develop and issue RFQ

                1.1.2.1.2   Work Package 3 – Design tower

Exhibit 1.3. Work breakdown structure example.

have been completed. Thus, the precedence order is complete design and engineering at which time the RFQ can be developed and issued. This then drives the RFQ response, analysis, and vendor selection, at which time an order can be placed for the tower.

The project team must develop the complete project task and activity precedence relationship and associated durations before they can build the project schedule.

Once the project schedule has been completed, the project team, usually using some project scheduling software such as Microsoft Project, Primavera, Spider, and so forth, will identify the project's critical path.

A project's critical path is described as the linkage between subsequent activities and tasks where if any one of these activities or tasks is delayed, as in starts late or finishes late, then the overall project completion date will occur later than originally planned. More fully defined, a critical path is "the longest path in a project network, because this path conveys information on how long it should take to complete the project . . ." (Monhor 2011, 615).

Project management literature describes the introduction of critical path planning and PERT as the beginning of project management as a discipline due to the application of management science (Carayannis, Kwak, and Anabari 2003, 1). Historically, the technique of developing a critical path is attributed to "... a joint effort in 1957 between the DuPont Company and Remington Rand Univac, to devise a technique to aid in the planning, scheduling, design and construction of large chemical plants.... [PERT] was devised in 1958 by the U.S. Navy's Special Projects Office, the consulting firm of Booz, Allen and Hamilton and the Lockheed Missile System Division" (Peterson 1965, 71). These processes support the approach that management decisions and techniques should be based on scientific principles. This understanding forms the foundation for the formalization of project management in the 21st century.

Returning to the history of projects and moving further into the modern era, we see a world which is consuming products and commodities manufactured across the globe or, as the process is sometimes referred to, globalization. Globalization occurs as entities in different countries interact on a commercial and economic base for various reasons, such as expanding their consumer base. Globalization becomes a driving force for organizations to build better products, faster, and at lower costs. The need to produce end products and services better, faster, and at lower costs is based on the need to be competitive with the global competition. No longer is a company just competing with firms within their geographical region or nationally but internationally as well. This international competition leads firms to leverage those capabilities which provide overall greater capabilities and functionality within their global setting. One approach, which has proven to be successful, is leveraging various project structures and project management principles. The project management discipline enhances the firm's ability to meet the drivers of developing new products at a faster rate, at a lower cost, and with increased quality.

Developing products and services faster, at lower cost and increased quality has specifically advanced due to the application of project management processes and techniques. As the International Project Management Association (IPMA) International Competence Baseline (ICB) states "The number of projects, programmes and portfolios is growing at an exponential pace, worldwide. In the past thirty years project management has been a discipline which has developed tremendously and increased in visibility" (Caupin et al. 2006, 2). With the development of critical path and PERT, the project management discipline continues to expand based on basic and applied research. The following sections provide a deeper view of the project management system as we see it today.

1.1.2 PROJECTS, PROGRAMS, AND PORTFOLIOS: WHAT DOES ALL THIS MEAN?

The previous section is written as if the reader has a full comprehension of what projects are and their various components. This section provides the reader a set of definitions, further background information, and a deeper understanding of what a project is as well as expands the overall project management view into project programs and portfolios.

To develop an understanding of what projects, project programs, and project portfolios are requires a clear definition and understanding of the contextual settings where these terms and project systems are applied. Science requires the use of exact definitions to ensure the removal of as much ambiguity in the study or experiment as possible. It is clear that confusion results when communication occurs in a context of seemingly common terms and phrases while the communicators have different understandings of the terminology. To minimize potential communication issues, this section provides the project management definitions which will apply throughout this book.

To start, we first look at the origin and definition of *project*. The word project is of "late Middle English.... [Where the] early senses of the verb were 'plan, devise; and cause' to move forward" (*Oxford Dictionary* 2013). From this we can discern that the verb project originally referred to an active, action-based process or a process which could be associated with making things happen and implementing change. So has the meaning changed over time?

A short review of recent project management literature identifies that various authors define project in different ways:

- "A plan or proposal; scheme. An undertaking requiring concerted effort" (*The American Heritage® Dictionary of the English Language* 2001).
- "… a time and cost constrained operation to realize a set of defined deliverables … up to quality standards and requirements" (Caupin et al. 2006).
- "… temporary endeavor undertaken to create a unique product, service, or result" (PMI 2013).
- "A time and cost restrained operation to realize a set of defined deliverables … up to quality standards and requirements" (GAPPS 2006, 6).

This set of definitions demonstrates how the term project may be defined depends on the author and his or her intent. This small example set supports the position, when analyzing the term projects, that "… although it is hard to find two identical definitions—all definitions revolve around a common center…." (Munk-Madsen n.d., 10). The common center is linked back to the late Middle English form of action and the process of implementing change. Thus, while individual authors provide varying definitions, the core concept of action and change remains a constant theme.

As such, we join the other authors and provide a definition grounded in action and change. For the remainder of this book, we define the word project as a structured sequence of events designed to deliver a distinctive product or service within a defined and constrained time and cost.

This definition can be universally applied to all projects regardless of specific contextual areas. Whether the project occurs during a basic research and development process, during applied research, within product development or application of services within a service organization, volunteer organizations, government entities, public sectors, and typical construction activities, this definition is applicable.

With a definition of what a project is, we turn our focus to what is a program within the project environment. In reviewing project management literature as we found with the word project, we identify that various authors use the words projects and programs interchangeably and that many definitions of the word programs also exists. The challenge with the inconsistency of using projects and programs interchangeably is that there is a distinct difference between what a project is and what a program is.

Since we have already defined a project, we can leverage this to define a program. Contextually, a program occurs when a set of related projects are performed which support a common company or organization strategic goal. These common projects are grouped together under a single management structure which is the program. The major differentiators between a project and program include the definition that a project is a singular initiative while a program involves multiple, organizational strategic, supporting projects. This set of strategic supporting projects is then grouped under a common management structure. This common management structure will include a project manager for each project and an overall program manager who will have the multiple project managers reporting to him or her.

As with projects, one finds programs in many different contextual settings. These settings generally include larger public and private entities rather than small government offices or smaller private organizations. This general differentiator is grounded in the fact that smaller organizational structures typically have less need, financial capabilities, and personnel resources to support multiple strategically aligned projects at the same time. Larger organizations, on the other hand, can and do find themselves in the position of having a common set of strategically aligned projects.

For this book we adopt the international standard definition of a program as "… generally a group of related projects and other activities aligned with strategic goals" (ISO 21500 2012, 6). Exhibit 1.4 shows the relationship between ranges of related projects which ultimately report to the program manager.

The last major grouping of projects is referred to as a portfolio of projects. ISO 21500 defines "A project portfolio [as a general] collection of projects and programmes and other work that are grouped together to

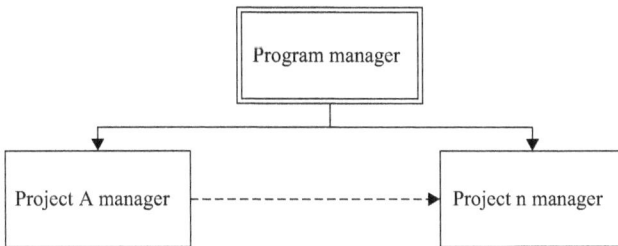

Exhibit 1.4. A program relationship chart.

facilitate the effective management of that work to meet strategic goals" (2012, 5). The differentiator between a program and a portfolio is that the program is a group of related projects which support a common strategic goal, while a portfolio is a collection of unrelated projects which are managed as a unit to achieve economies of scale. This is a fine line distinction but, within a program all projects have a relationship with each other, while in portfolios the individual projects have no distinct relationship other than supporting the company's or organization's strategic objective. Thus, the differentiating question is whether a set of projects grouped under a common management structure is a program, or a portfolio is the grouped set of projects specifically related to each other, or are they a set which just supports a common strategy such as developing a set of new products and services?

Another way of looking at the difference between a program and portfolio includes how the corporate benefits are obtained. Within a properly structured and coordinated program, the maximum benefits are achieved when all associated projects are completed. There is a direct synergy between each project's outcomes and the overall program. Portfolio project benefits occur as each portfolio-associated project is completed. While the corporation's overall benefit continues to increase as each project is accomplished there is no synergy, other than portfolio oversight, that is achieved between the individual portfolio projects.

For this book we adopt the ISO 21500 portfolio definition as presented in the previous paragraph. As with programs, portfolios occur in most organizational contexts. The frequency of occurrence in small government organizations or private firms is less than what occurs in larger public and private organizations. The frequency of occurrence is driven by the need for multiple projects to occur at the same time as well as the availability of financial and personnel resources to support the multiple projects.

Exhibit 1.5 depicts a set of unrelated projects which report to a single portfolio manager. While this structure is very similar to Exhibit 1.4, as

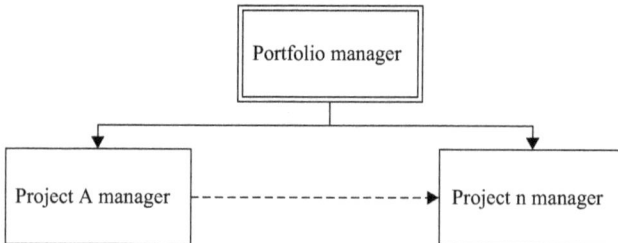

Exhibit 1.5. A portfolio relationship chart.

noted earlier, the key distinction is program projects are related to each other, while portfolio projects are not.

To sum up the difference between a project, a program, and a portfolio; a project is a single unique initiative while programs and portfolios encompass multiple projects. A program is a set of related projects which support a common organizational objective. There is synergy between the projects which derives a greater corporate benefit when all are completed. A portfolio is different in that it includes a set of unrelated projects which are managed as a unit by a common portfolio management structure which supports a key overarching company or organization strategy.

1.1.3 PROJECT MANAGEMENT, PROGRAM MANAGEMENT, AND PORTFOLIO MANAGEMENT DEFINED

To be successful projects, programs and portfolios must be *managed* using a set of tools, techniques, processes, and procedures, that is, the project management structure. Conveniently, these structures are named project management, program management, and portfolio management. Unfortunately, the project management literature, as with the definition of projects, programs, and portfolios, lacks consistency and agreement on the final definition. A review of the project management literature identifies that the authors apply project management and program management interchangeably as if they are synonyms, which they are not.

Before we delve into defining and briefly discussing the set of project and program management structures, within the project environments, we will briefly review what management is.

Two of the most frequently referenced management writers are Henrie Fayol and Frederick Taylor. As Fells points out, "Although Taylor's work is sometimes compared with Fayol's, it is important to realize that the focus of each is quite different …. Fayol viewed management from the executive perspective while Taylor focused on the other end" (2000, 345). As seen from the executive perspective, management is inclusive of planning, organizing, coordinating, controlling, monitoring, and adjusting. This can be viewed as a circular, continuous improvement process as shown in Exhibit 1.6.

Continuous improvement occurs as management monitors the process and compares the processes and results to the planned objectives. Based on the comparison and analysis, the organization can make adjustments as required. Over time the organization continues to refine and adjust, as required, to improve the overall process outcomes and objectives.

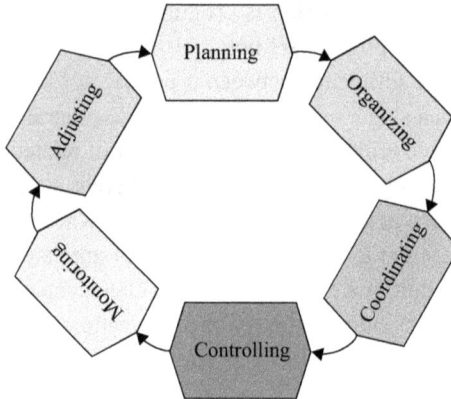

Exhibit 1.6. The management continuous improvement circle.

ISO 21500 defines the management process slightly different as initiating, planning, implementing, controlling, and closing (2012, 12). While it appears that ISO 21500 defines the management process slightly different, in reality, it is in alignment with the broader management definition.

From this book's perspective, project, program, and portfolio management falls within the overarching management discipline. Comparing various project and general management studies it is determined that the primary responsibilities of these positions are the processes of planning, organizing, coordinating, controlling, monitoring, and adjusting within the project, program, or portfolio. This position is supported by entities such as the Global Alliance for Project Performance Standards (GAPPS). From GAPPS documentation we find that "Project managers, program managers, and portfolio managers are expected to produce essentially the same results—outputs and outcomes that are acceptable to relevant stakeholders" (2006, 4). The acceptable outputs and outcomes of the relevant stakeholders are achieved through the management process of defining or planning, organizing, implementing or coordinating, controlling, monitoring, and adjusting the sequence of events which constitute a project.

While general management and the various project management positions have similar roles and responsibilities, there are distinct differences between general management and project, program, and portfolio management positions. The primary differences between the disciplines are that decisions and processes within projects are time constrained and involve one of a kind initiative.

The combined features of an explicit time constraint and unique, one of a kind, initiative result in a different level of complexity between general management and project management as well as different complexity

$$Complexity_{program(portfolio)} \equiv \sqrt{\sum_{1}^{n} Complexity_{program}^{2}}$$

Exhibit 1.7. Complexity formula.

levels between the three *project* management functions. One can view program and portfolio management as efforts with greater complexity than a single project management effort. For every project within the program or portfolio its individual complexity factors combine into the program or portfolio management complexity. One could view program and portfolio complexity as approximately equal to the square root of the sum of squares of the individual project complexity (Exhibit 1.7). Part of the approximation occurs due to some complexity reduction which occurs through the program or portfolio's management structure.

In summary, there are distinct differences between project, program, and portfolio structures and complexity levels. Projects involve a single initiative; programs are inclusive of a set of related projects while portfolios are sets of unrelated projects. The number and complexity of individual projects combine to increase the overall complexity of programs and portfolios as well. We also identified that there are distinct differences between the general management, project management, program management, and portfolio management disciplines' roles and responsibilities. While each of these disciplines shares common areas, management within the project environment is different than general management as projects are different from day-to-day operations.

1.1.4   TYPES OF PROJECT TEAMS

Projects are implemented by a project team. The number of people, technical skill sets, and processes required for each project is as unique to each project as the project itself is unique from all other projects. In looking at the project continuum complexity scale, project teams can be very small and involve a single technical skill on one end of the continuum scale to projects which include thousands of workers with an extensive range of skills required and a depth of different processes on the other end of this scale. As an example, the development of this book project, to a point it was ready for printing, involved a small group of people who included the author, proof-reader, and editor and took two years. Yet, the Trans-Alaska Pipeline took three years to build, required tens of thousands of workers with scores of different disciplines and skill sets.

To facilitate a continuum scale of project sizes and a wide variety of project teams, several different project team structures have been developed and applied around the world. The most frequently cited and discussed project management team structures include (a) the dedicated or projectized project team and (b) the matrix project team. These types of teams tend to be associated with *traditional* organizations and project environments where the team members work within the same geographic area.

Dedicated or projectized project teams (from here on, dedicated teams) are groups where all members are assigned to the project full time. In this structure, the team members report to one boss, the project manager, and have one assignment, work on the specific project. Exhibit 1.8 demonstrates such a team. The individual team members report to their respective discipline lead such as the engineers report to the engineering lead. The various discipline leads report to the project manager. A clear line of authority exists from the project manager to the individual contributors.

Dedicated project teams follow the traditional vertical organizational structure. In the vertical structure, each discipline forms a unit which is sometimes metaphorically referred to as a silo. Within silos, it is only at the senior levels where direct interaction between the disciplines occurs. This is a potentially negative issue which occurs in a dedicated project team environment—that is, the individual team members fail to interact and exchange information directly but rely on information transfers at the senior levels.

There are several positive aspects of a dedicated project team which include (a) the team has one work objective which is completing the project, (b) the project team members have a single boss who is the project manager, (c) the team can become cohesive in a shorter period of time as they work together on a full-time basis, and (d) a project-specific culture can form.

While there are several benefits to a dedicated project structure, there are negative aspects as well. A full-time, dedicated project team may actually cost the project more over the course of time as the outcome of not fully utilizing all skill sets for all hours of every work day. There is also the negative aspect of where to transition the team members when the project ends. If there are other projects that the various team members can rotate to or if their operational job is still available, this helps minimize this negative impact.

Successful dedicated project teams have a shared culture which defines how they interact. Various visible cultural attributes include a common language, ceremonies for success and failures, methods of problem solving and decision making, as well as how the team's work settings

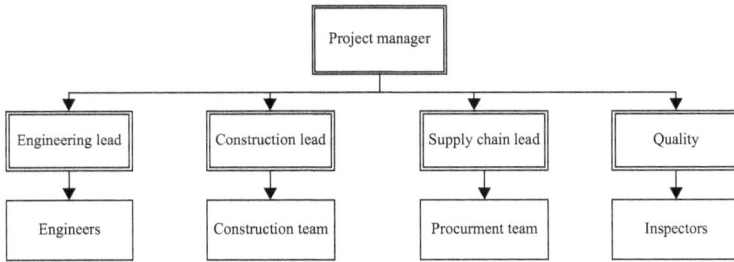

Exhibit 1.8. The dedicated project team structure.

are designed. Development of what the project team accepts within the social environment occurs more easily with dedicated project teams, especially if they share a common work location. This is driven by the fact that the team is forced into common, daily interactions, and social context. What is and is not acceptable can quickly become apparent and be integrated throughout the team.

As noted previously, matrix project teams are another very popular project team structure. Matrix project teams occur when the company *loans* the project resources for a percentage of their scheduled work time. The matrix project team structure creates the two-boss situation where the team members directly report to both their functional manager and the project manager.

Exhibit 1.9 is one method of showing how this two-boss linkage may appear within an organization. The left column reflects the firm's operational organizations of engineering, construction, supply chain, and quality departments. Across the top is the project manager and below this are the project functions of engineering, fabrication, procurement, and quality. As stated in the previous paragraph, the operations side of the firm loans the project specific individuals to assist in the project. The individual team members retain their reporting relationship to their operation lead as well as to the project management lead.

Positive attributes of the matrix organization may include (a) better utilization of resources, (b) lower cost to the project, (c) a means to enhance project team member skills as well as (d) job satisfaction.

Better utilization of the assigned resources is the opportunity to use a technical resource on both operational issues and on more than one project during the same time period. The allocation of work is dependent on the person's available time, specific assignment requirements, and work assignment complexity. The key attribute is that with proper management and coordination of the technical resources work and time commitments, their effectiveness and efficiency can be maximized across different needs.

Exhibit 1.9. The matrix reporting structure sample.

Operation lead	Discipline	Project manager			
		Engineer-ing	Fabrica-tion	Procure-ment	Qual-ity
Engineering lead	Mechanical	X			
	Electrical	X			
	Civil		X		
Construction lead	Mechanical		X		
	Electrical		X		
	Civil		X		
Supply chain lead				X	
Quality lead	Mechanical				X
	Electrical				X
	Civil				X

Lower cost to the project is related to the better utilization of the assigned individual's time factor as well. As the technical resource is leveraged across different efforts, the project team is not required to pay for a full-time equivalent. Conceptually, the project only pays for the time applied to the project which would be very focused with specific deliverables. This would result in obtaining the activity output at a lower cost than a full-time assigned resource may cost.

The potential for project team member skill enhancements and job satisfaction comes about as the team member is exposed to new and unique challenges which occur within the project context. As projects are unique endeavors, the team members could be exposed to challenges which they may never see in a day-in-day-out operational support role. This is in alignment with Herzberg's motivation theory, that is, job satisfaction of increasing job status through increased opportunities and responsibilities that help to increase the individual's status within the organization. Being part of a successful, complex, and high-risk project provides a linkage to these factors.

There are some potential negative sides to the matrix organization as well. One issue is the dual reporting structure where the team member reports to both an operational lead as well as the project manager. This is referred to as a dual-boss situation. Anytime you have two bosses there is the potential for conflicting directions, needs, and requirements. Generally, the project management research identifies that "Direct authority over personnel tends to rest with the functional manager ... [who has] control over context ... specifically, including those associated with technical specialties, while including career development and growth ... [while] project managers are focused on the project, with little direct authority over the project team members...." (Dunn 2001, 5).

The two-boss environment sets up a not uncommon win–lose situation. In this case there will be a winner, such as operational needs *win* by obtaining what the operation requires, and there is a loser, such as the project *loses* when the project efforts are negatively impacted. This can be related to Herzberg's dissatisfaction job environment factors of supervision and working conditions. The two bosses create confusion and competition between the various supervision and working conditions. In the end, the resources can be negatively impacted as they feel like the working conditions are less than ideal, there is confusion on what they are to do, and they see dissatisfied managers as well.

Other common negative outcomes of the two-boss issue are decision distress and power struggle. What occurs is that either or both managers feel that they must make all decisions without support from the other, or they feel like they have no say in the decision process, or they view the other boss's action as trying to push the tough decisions onto them to avoid a difficult situation. The different managers may also be working against each other in an effort to place themselves in a better political position for the next promotion, pay raise, or premier project. The political cross purposes can result in severe negative impacts to the project as the focus is on improving the manager's position rather than optimizing the operational and project work efforts. Under these various situations, the project team members are placed in a state of confusion as to who is really in charge, what should be their priorities, and whom should their primary allegiance be to. As such, in any of these environments, productivity, quality, and performance capabilities are likely to decline and job satisfaction suffers.

To minimize and hopefully avoid the potential negative project results, senior management must be engaged and involved in the project. A key senior management function is to monitor the two-boss state and address these issues. Without this oversight, the probability of project team success is greatly reduced.

Another potential negative impact of the matrix project structure is poor communications. While, conceptually, communications should improve as the project manager and organizational managers learn to communicate with each other to balance the competing requirements, research has not fully validated this outcome. Unfortunately, a frequent decision process is capitulating to the "… low-est-common-denominator political compromise. Top managers were spending more time in meetings…. Gamesmanship and political jock-eying were widespread" (Peters 1979, 15). In essence, this research indi-cates that communications suffered and decisions were being made based on lowest-common-denominator, gamesmanship, and political jockeying versus what would be the optimum for the organization and the broader team. Again, this situation highlights the need for senior management engagement to ensure these types of issues are addressed to increase the project's efficiency and effectiveness.

Successful matrix project teams must have a shared culture which defines how they interact even if they are only assigned to the team part time. That is, the project team members must still develop a shared com-mon language, ceremonies for success and failures, methods of problem solving and decision making as well as how the team's work settings are designed. Development of what the project team accepts within the social environment may take longer to occur when project team members are not assigned to the project team full time and are physically separated from many of the project team members on a daily basis and have another—possibly diametrically different—culture in their functional or operational role.

When analyzing various matrix organizations, one will encounter a range of types which are labeled from weak to strong. A weak matrix orga-nization is often described as one where the project manager is more of a coordinator than a manager or a leader. In the weak matrix, the functional manager retains virtually all authority over the team members and decides what, when, and where the team members will work. In a strong matrix project structure, the project manager has full responsibility for delivery of the project and a high level of authority over the individual team mem-bers. In this structure, the functional project manager is consulted as to work schedules, other organizational needs, and processes but the final decision rests with the project manager.

In general, in dedicated and matrix project teams everyone has a com-mon objective of making the project successful. This common objective supports a cohesive project culture, project vision, mission, and goal. Very much like systems engineering concepts where the output of the system is greater than the sum of the individual parts, high performing dedicated or matrix project team's capabilities are greater than the sum of the capa-bilities of the individuals' team members. In support of this position, sys-tems engineering identifies how the project outcome is greater than the sum of the inputs. This results from the system's positive synergy which

"... is derived from the Greek work sunergos: 'working together' .... The essence of positive synergy ... [is] The whole is greater than the sum of the parts" (Gray and Larson 2008, 349).

While dedicated and matrix project team structures receive the most research attention, the literature also identifies the ad hoc project team structure as existing with projects. "The dynamic life cycles of systems place demands on the organization which are different from those traditionally felt by managers. These demands have resulted in the creation of ad hoc 'teams' or 'Projects' as organizational devices for coping with these new phenomena" (Cleland and King 1983, 245).

Ad hoc project teams are different than the dedicated or matrix structures in the way the team members approach the project. These project teams are also characterized differently than the systems concept as well as the dedicated or matrix organizations in that the shared project end objective, vision, mission, and goals are lacking and the sum ad hoc team output may not exceed the sum or the individual parts.

> In many ways an ad hoc team is a collection of individuals who have each put their commitment to themselves above their commitment to the team. That mentality threatens to negate all the hard work project managers have put into learning how to build successful, cohesive and empowered teams. As one project manager laments: "The only person who faces career failure in this team is me!" (Bushell 2004)

Ad hoc project team members' commitment can be mapped out with the number one objective being their own personal agenda. That is, their greatest interest is their own advancement and how to maximize their outcomes. The second level commitment may be to the functional team they deal with on a day-in-day-out basis. This secondary commitment may be driven by the view that supporting their functional team's success supports their number one self-commitment driver. Commitment to their overall organization may come in third on the list. Again, this is probably centered on the idea that they need to be associated with a successful organization to foster their own success further. Any commitment to the project team falls way down the list if at all.

Ad hoc project teams are faced with many potential negative issues such as:

- Lack of trust—everyone is looking out for themselves and not for each other.

- Increased internal conflict—with no cohesive team and each person looking out only for his or her own needs, conflict between team members is highly probable.
- Dynamic participation—ad hoc teams may experience greater turnover in team members as the individuals find *better* opportunities to pursue and leave the team.

With expanding globalization, a different type of project team has emerged. These project teams are not bound by the team members' geographic location or physical constraints such as a dedicated office. This new global project team structure is often referred to as a virtual project team. In this case, virtual only refers to the fact that the project team is never located at the same physical location.

Virtual teams consist of individuals working toward a common goal yet, generally they are not located in the same physical site. Their primary interaction occurs through technology such as telephone, Internet telecommunications, e-mail, and video teleconferencing. "However, virtual teams can take on a variety of forms, such as groups containing some co-located members and some distributed members...." (Webster and Wong 2008, 41). Assignment to a virtual team may be as a dedicated team member, part of a matrix structure, or even an ad hoc team structure as well.

The key differentiator between the more traditional project team structures previously discussed and the virtual project team is the fact that the virtual team members may never be physically in the same place at the same time. All interactions must occur through the use of technology such as telephone conference calls, e-mails, text messages, shared and collaborative Internet and intranet websites, and video teleconferences.

Virtual project teams, besides the challenges of interacting primarily through technology, also face the difficulty in establishing a project culture. "Colocation, or physical proximity more generally, is said to reinforce social similarity, shared values, and expectations ...." (Jarvenpaa and Leidner 1999, 792), that is, team culture. When the close physical proximity does not exist, developing and sustaining a shared project culture is harder to achieve and may never occur.

While each of the various project team structures has positive and negative aspects, a common theme across all of them is project team culture. The literature is clear that having the proper project team culture contributes to the project's success. "So, in essence, culture influences how each of the organizational components ... is designed. In turn, the design shapes what behavior is emphasized and rewarded, and which modes of operating are discouraged and constrained" (Smith 2010, 22). Yet, each of the team structures has varying opportunities to develop a cohesive culture, thereby requiring different approaches. Later chapters in this book expand on this aspect of the project team in much greater detail. The next section looks at project management processes to introduce the reader to project management methods, tools, and techniques as a means to continue with building a common vocabulary.

1.1.5 PROJECT MANAGEMENT PROCESSES

In this section, we provide a general overview of the processes involved in planning, executing, monitoring, controlling, modifying, and closing out a project. The project management literature identifies that this is a process which includes a team working together toward a common goal. It involves the development of a variety of plans, leading and managing the team toward the end result, all within financial, time, and quality constraints.

Before project management even begins, someone has to identify a need or want. From this kernel grows the overall project until an end is obtained. This process forms the project life cycle.

Life cycle is defined as the series of sequential steps, processes, or phases which start with the idea and end when the project completes. From the literature it is clear that there are many different life cycles which have a variety of phases as well as different names for each of the phases. The literature is clear that there is no universal practitioner, researcher, or standard organization acceptance of how many phases exist or even what the various phases should be called (Faulcombridge and Ryan 2002, 5). The Project Management Institute's (PMI®) standard, Project Management Body of Knowledge (PMBOK), identifies several different life cycles to include (a) predictive life cycles, (b) iterative and incremental life cycles, and (c) adaptive life cycles (PMI 2013a).

Predictive life cycles tend to be associated with projects that are more structured and where the project team views change as an exception versus a normal routine. While the exact phases of a predictive life cycle vary according to the project needs, project team composition, industry sector, and company standards and culture in general, the predictive life cycle phases generally or typically include (a) concept, (b) preliminary design, (c) detailed design, (d) implementation, and (e) closeout.

Each of these general phases can be and is often broken down into finer delineation of work efforts in a decomposition process of providing greater detail within the overall process. As an example, in the concept phase, several different activities occur such as first identification and documentation of the need or want. The team then performs an analysis as to final output performance requirements, potential solutions and alternatives, general ideas of cost and schedule requirements, as well as resource requirements. The team will also analyze how the potential end product or service would fit within the organization's strategic plan and current business structure. A key element of each of these efforts would be a document or series of documents which provides input into the next phase.

One way to view this life cycle is erection of a physical structure. The concept phase is the development of the structure's foundation from which all else is supported. If sufficient time, attention to detail, and focus is applied, the foundation will support the structure. If the concept phase is rushed or lacks sufficient detail and analysis the resulting structure may fail from its own weight.

There are many different predictive life cycle types in use throughout the world. Some of these include the waterfall, the spiral, and the rapid application development or evolution.

Waterfall—this life cycle is most commonly used and follows seven phases:

1. Concept development
2. Requirements development
3. Preliminary design
4. Detailed design
5. Implementation
6. Testing
7. Closeout

A key feature of the waterfall approach is that each phase has feedback loops to the phase immediately preceding the current phase. Exhibit 1.10 demonstrates this feedback loop. In the original format, the waterfall process proceeded from one phase to the other with no feedback (Exhibit 1.11). A critical outcome of the feedback process is that it helps tie the phases together and provides improved project management processes while minimizing the potential for endless loops back to earlier decisions.

Spiral—this life cycle approach is often used in software development areas. In essence, the overall project is separated into a series of smaller projects with a common phased approach. There are five general spiral phases:

1. Identification of this iterations objectives
2. Identification and resolution of the risks

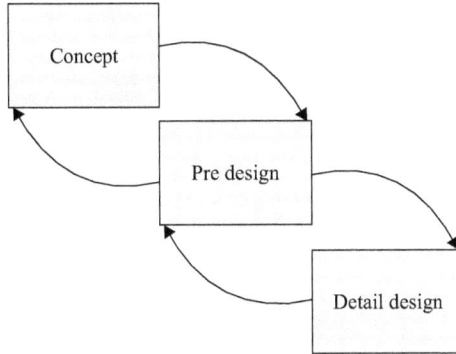

Exhibit 1.10. The waterfall with feedback.

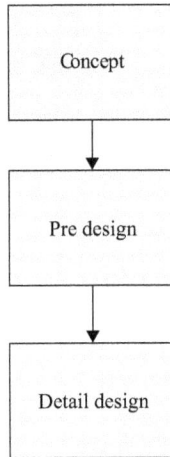

Exhibit 1.11. The waterfall with no feedback.

3. Implement the waterfall life cycle phases
4. Plan the next iteration
5. Repeat

The spiral life cycle approach first appeared in 1998 and is attributed to Barry W. Boehm. "The major distinguishing feature of the spiral models is that it creates a *risk-driven* approach to the software process rather than a primarily *document-driven* or *code-driven* process" (Boehm 1998, 61). Exhibit 1.12 provides a general view of the spiral life cycle process. As shown, the project goes through the series of phases several times until the project is complete.

Cumulative costs

Determine objectives,
alternatives,
constraints

Evaluate alternatives,
identify, resolve risks

Review/commitment

Plan next phase

Develop, verify next-level
product

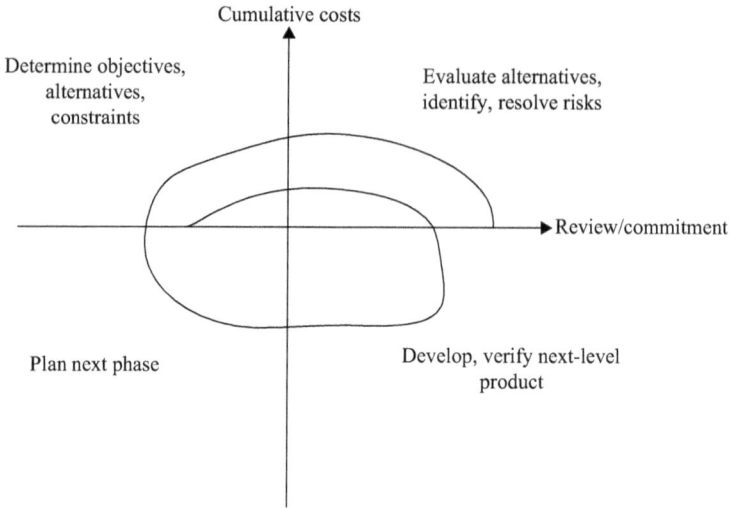

Exhibit 1.12. The spiral.

Rapid application development or evolution—in this life cycle, the project evolves through five phases:

1. Concept
2. Analysis requirements
3. Design prototype
4. Build prototype
5. Review prototype with project sponsor
6. Repeat from step 2

The positive view of this life cycle is that (a) it keeps the customer engaged with the process and (b) it provides continual feedback to the customer as well as the project team of the project's overall status.

The negative aspects of this effort, (a) the lack of a defined schedule, (b) associated costs, (c) the potential that numerous prototypes can be built, and (d) final product deliverable quality requirements, which occurred at project initiation may not be that which the project ends with. That is, project quality can become a moving target.

Across the various industries, you will encounter various versions of these basic predictive life cycles. The versions will be based on the industry, how the company wants to perform projects, and the project manager's preference.

For projects where change is a normal event and accepted by the project team, clients, and stakeholders, adaptive life cycle approaches are often used. The following identifies some of the adaptive life cycle models discussed in the project management literature:

- Adaptive Software Development (ASD): Mission-driven, component-based, iterative cycles, time boxed cycles, risk-driven, and change-tolerant.
- Extreme Programming (XP): Teams of developers, managers, and users; programming done in iterative process, collective code ownership.
- Agile and SCRUM: Similar to the above-mentioned adaptive life cycle models with iterations called *sprints* that typically last one week to 30 days with defined functionality to be achieved in each sprint; active management role throughout (Archibald, Di Filippo, and Di Filippo 2013, 6).

There are several key elements of the adaptive life cycle. First, the life cycle consists of several iterative cycles where change can be introduced at any time. Second, the project team plans for and adapts to changes that occur throughout the project as the team expects and plans for changes. The project life cycle is one of learning and adapting where current insufficient information is expanded on during the next iteration and the project scope is further defined.

Adaptive life cycles also provide the opportunity for incremental output releases. As an iterative cycle is completed the project team can release that version to the client (user) for further verification and testing. These interim releases support the clients' developing an understanding of the current capabilities and develop a better vision of what they want the final product to be.

While all project life cycles must take into consideration that change may occur, adaptive life cycle projects are specifically designed to incorporate almost continuous change processes while predictive life cycle processes view scope change as an exception which must go through a rigorous change management process before it is accepted.

The previous discussion highlights the two principal life cycle categories most often seen in practice and discussed in the research as well as some of the specific life cycle types which are found within these principal classifications. As stated earlier, which project life cycle is utilized is dependent on many factors such as the type of project, that is, construction or software development, the company's preferred or standard approach as well as its project management maturity, the project manager's preference, contractual requirements, the company's culture, or all these aspects.

Organizations have a culture and that culture has an impact and influence on the success of the project and its ultimate performance (PMI 2013a, 20). While further details of culture are presented in the following chapters, it is acknowledged that each organization has its own culture and that within organizations project-specific subcultures may exist as well. These unique cultures occur through the interaction of the team developing norms and behaviors, for example, "over time we internalize these behavior[s] ... and we no longer have to think about what to do or say . . . we just know" (Storti 1994, 4). Therefore, a project team may implement a specific life cycle because that is the way it is performed within their organization rather than on an extensive review of the potential options and alternatives.

Keeping in mind that the exact life cycle approach is dependent on many factors, all projects should follow a sequence of distinct events which lead from the conceptual idea, want, or need to project conclusion. Often these series of events are repeated during many or all of the life cycle phases and, as such, are not life cycle phases themselves but major project activities with distinct outputs and potentially unique outputs and outcomes.

Outputs are defined as the end result of the process which may be a product or service (PMI 2013a, 548). Outcomes are distinct events within the project which contribute to the final output. As an example, during the project specific sets of documentation are developed to document scheduled versus actual spend cost reports. These are outcomes of the project rather than outputs. Outputs are the palpable, measurable, quantifiable project results while outcomes are the completed processes that occur throughout the project.

The first effort of any project team and its associated stakeholders is the need to produce the first outcome—the project SOW. This occurs in what is generally referred to as the project initiating phase. Those associated with the project at that time formalize the original project idea into a document which includes the following key elements—SOW:

1. Project SOW statement: The SOW statement can be as short as a single sentence or a very lengthy description of what the project is to accomplish.
2. Strategic alignment statement or justification: The strategic alignment or justification portion of this document provides the clear linkage between the company's strategic plan, organization needs, and how the final output or outputs will integrate within the overall company.

3. Identification of any known restrictions or constraints: During the project initiating phase the stakeholders, project team members, and others may identify items which may restrict or constrain how the project is implemented, testing methodologies, technology limits, and process limits, and so forth. Defining exactly what restrictions or constraints are already known helps the project team plan the overall project.

4. Identification of the project output or outputs: The SOW will provide a clear definition of all output(s) that it will deliver.

5. Known acceptance metrics: The SOW must also detail what metrics will be used to accept final delivery of the project output. While the SOW will not include specific tests, procedures, or processes, it will define the core functionality of the output and how it is measured.

With the SOW in hand, the project team can proceed forward in developing the overall project management plan. This plan is fully inclusive and detailed on what processes are utilized and how they are managed. Core sections of the project management plan typically include the following:

1. Work breakdown structure (WBS). A WBS is a detailed task listing of all main efforts within the overall project. While the level of detail provided in a WBS varies from project to project, the level of task breakdown must be sufficient so that the project team can plan the project and fully understand the major tasks, their linkages, and interdependencies.

2. Project schedule. The project schedule is a detailed network map of the full project. To develop the project schedule, the project team takes the WBS and subdivides it further into work packages that have definitive time durations, identification of predecessor and successor relationships, as well as identified resource requirements. With this level of detail, a project network diagram is developed, that is, the project schedule.

3. Cost detail or spend plan. The cost detail or spend plan is built from the overall project schedule. This portion of the project management plan identifies items such as the anticipated spend rate and major purchase requirements.

4. Communications plan. All projects require internal and external communications. The communications plan details to whom, how, what, by what method, and frequency these interchanges will occur.

5. Risk management plan. Projects are unique endeavors which always carry a level of risk. The risk management plan identifies

the known risks, what indicators or information will signal that the risk is about to happen, that is the risk trigger, and the project team's mitigation response plan for each risk. The risk management plan will also detail the frequency updates and how unexpected risks will be managed.

6. Project team management plan or human resource management plan. Depending on the duration of the project and frequency of project personnel, turnover is to be expected and planned for. The management plan details how these changes will be managed as well as each team member's role, responsibility, and level of authority.

7. Quality plan. To ensure that the project meets the output requirements, a quality plan is required. This plan will define both the quality assurance and quality compliance processes.

8. Project change control process. As projects are subject to change and change generally impacts any of or the entire major attributes of the project schedule, project cost, project output, and project quality requirements, the project team must document how it will handle change, which is the change control process. This part of the project plan provides representative samples of the change request form, identification of who must review or approve the request, how the change is communicated, as well as how updates to the impacted documentation are handled.

9. Project procurement plan. Many projects require procurement of tools, material, supplies, software, or assembled units. The procurement plan details how these purchases occur, each project team member's level of procurement authority, as well as procurement approval processes. The plan will also detail how material is received, any inspection requirements, defective material return and repair process, and disposal of surplus items. The procurement plan may also detail where project material is temporarily stored, checked into and out of the storage area, and inventory processes.

10. Project monitoring and control plan. A key project team activity is monitoring and controlling the project. This section of the project plan will provide details on how all the other project plan major actions will be monitored, the essential metrics that are tracked, as well as what processes will be used to report the project status. The monitoring and control plan provides direct linkage between the monitoring processes and how project corrections are implemented and communicated.

11. Project closeout. The project plan identifies what the output(s) are and what the acceptance criteria or metrics will be. Included in the closeout plan is how the project outcomes, for example documentation, will be disposed of, transferred to the company record storage, or integrated into company-specific processes. The closeout plan also defines how the project team will implement the system final testing processes, how to deal with any final issues, as well as how the project team members will be transitioned to either other projects, returned to their original departments, or let go.

When the project management plan is developed, reviewed, and accepted by the project team, stakeholder, and required company personnel, this signals establishment of the project baseline. This baseline forms the metric from which the project will be implemented, monitored, and controlled.

The process of performing the actual project is labeled project implementation. Project implementation involves executing the project management plan and modifying the plan as the need arises. Throughout the implementation phase, the project team will develop various project outcomes such as time and schedule earned value analysis reports, processing of any change request which may occur, as well as updates to all the project management subplans discussed in the preceding paragraphs. Unless the project management plan and schedule identify intermediary or interim outputs the project team will be producing outcomes, not outputs, throughput the implementation phase.

An essential project implementation process is the continuous monitoring and control function as discussed earlier. Throughout the project implementation, the project team members will execute the monitoring and control plan to ensure the project team, stakeholders, and other interested entities know the project status, identification of any risks that may have occurred, and details of any recovery plan that has been developed. Monitoring and control starts when implementation starts and ends with project closure.

All projects come to an end. The project has delivered the full SOW, some modified SOW, or the project was canceled before completion for any number of reasons. The project management team's final role is to ensure that all closeout documentation is complete and appropriately dispersed and the project team dismantled according to the plan.

In summary, the project management process starts with an idea, want, or need and culminates with proper disposal of all outcomes, delivery of the output(s), and dismantling of the project team. Accomplishing this full project life cycle begins with the development of a detailed project management plan, implementation of the plan, monitoring and controlling of the process, as well as assurance that all quality requirements are met through appropriate levels of testing and validation efforts.

At this point, the reader should have a firm grasp on what projects, programs, and portfolios are as well as a standard set of definitions for general project terminology. To help delineate the uniqueness of projects, the next section discusses and compares the difference between operation management, engineering management, and project management.

1.1.6 OPERATION MANAGEMENT, ENGINEERING MANAGEMENT, AND PROJECT MANAGEMENT— WHAT ARE THE DIFFERENCES BETWEEN THESE ORGANIZATIONAL ROLES?

If one spends a little time reviewing a bookseller's business section, be it online or at the neighborhood store, one will see book titles which include the names of operation management, engineering management, and project management. A quick review also identifies that there is a "… traditional view of management … [in] that it is a process concerned with achievement of objectives" (Cleland and King 1983, 10). While teaching these discipline courses, the author has been asked to explain what the difference is between these disciplines. This section addresses that question and helps to focus the remainder of the book on project management.

We start this discussion with understanding and defining OM. We then move to discussing engineering management and then we end this section with further discussion on project management, its role and responsibility.

To define OM, one must first understand and define what operations are. A common OM theme is that operations are a transformation process where raw material is transformed into a different state with added value (Russell and Taylor 2009). With this idea then, operations can be defined as "… activities that relate to the creation of goods and services through the transformation of inputs to outputs" (Heizer and Render 2006, 4).

Expanding on these definitions, operations is the ongoing and recurring processes of taking raw material, subcomponents, pieces, parts, or subassemblies, then adding value to the inputs, through some transformation process, to produce a desired output. Adding value to the inputs is the transformation process that the literature refers to. As a very basic example, when you purchase that hot cup of coffee from the barista, the operation process takes the input, in this case the roasted coffee bean, grinds the bean, and pours hot water over the ground beans to produce a cup of hot coffee. The value added steps, assuming that the barista purchases roasted coffee beans, are the steps of grinding the coffee beans, pouring hot water over the resulting ground beans, pouring the hot coffee into a cup, and presenting it to you, the client. It is a continuous process which has several different options but the basic process remains the same for each cup of coffee.

Exhibit 1.13 provides a general overview of this process. As shown, operations consist of some form of raw material input. In the example above, the raw inputs are the coffee bean and water. These inputs proceed through a transformation or value added process such as pouring hot water through the ground coffee beans. The output of the transformation process is the intended product or service. That is the hot cup of coffee.

The key to understanding what operations are is the aspect of repetitive systems. That is, the system performs essentially the same process for each sequence, day in and day out. The system receives the same raw input material and then produces essentially the same output by following a repetitive transformation process. This is not to imply that the product output cannot or does not change but that the different versions of the output are designed and implemented as an ongoing process rather than unique events. This is operations.

As operations are a continuous transformation or value added process, then, what is operations management? "Operations management (OM) [is] the business function that plans, organizes, coordinates, and controls the resources needed to produce a company's goods and services" (Reid and Sanders 2005, 3). From this we establish the operations manager's role as the company's executive who is responsible for the system which receives the raw input, adds values through some transformation process, and produces the intended output.

If operations managers are responsible for the full transformation system processes, then what is engineering management and how is it unique from operation management? The American Society for Engineering Management (ASEM) defines engineering management as "... an art

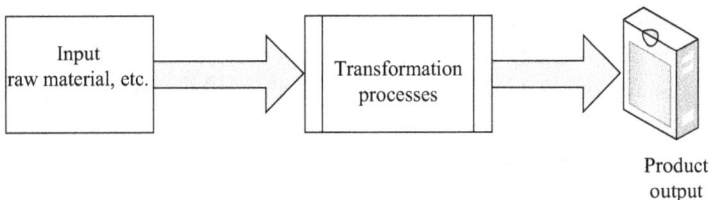

Exhibit 1.13. The transformation process.

and science of planning, organizing, allocating resources, and directing and controlling activities that have a technological component" (Shah 2012, 3). When analyzing engineering management university degree requirements literature it identifies that "… the most common descriptive terms for this degree program were: engineering, management, systems, project management, people, business, problem solving, organizations, cost/finance, communication and manufacturing/production" (Elrod, Rasnic, and Daughton 2007). The literature also states that there is a need "… for people to be educated in both technology and management in order to better manage the increasingly large and complex industrial and government projects (Pence and Rowe 2012, 46).

From these literature references it appears that there is overlap between OM and engineering management. Therefore, it is safe to assume that there would be an overlap between OM and engineering management definitions as well. The key definition differentiator is that operations managers are focused on the overall transformation system view while engineering managers' focus on the technological component of the transformation process.

Thus, engineering management focuses on the merging of general management process as applied to the operations and production transformation technology system component. This text adopts a definition of engineering manager as the company's executive who is responsible for the merging of technology and general management principals to ensure an efficient and effective transformation process which receives the raw input, adds values, and produces the intended output. The essential differentiating point is that engineering managers' primary focus is the technology function of the transformation processes not, necessarily, the overall transformation process.

The next consideration is what differentiates project management from OM and engineering management? As stated previously, management is traditionally viewed as the process of objective achievement while project management literature states that projects are intended to deliver specific outcomes or objectives as well. While, on the surface, there appears to be a similarity between OM and project management, the project management literature is clear that a distinct difference exists between them as well as between project management and engineering management.

To add specificity to what differentiates project management and the other management disciplines, we look to the various published project management definitions.

PMI defines "Project management [as] the application of knowledge, skills, tools, and techniques to project activities to meet the project requirements" (2013a, 5).

ISO 21500:2012(e) defines project management as "... the application of methods, tools, techniques and competencies to a project .... Project Management is performed through processes" (4).

The project management system is all encompassing. It includes a definitive organizational structure, defined procedures, data and information processing, and unique methods and tools. "The project management system provides for integrative *planning and control*" (Nicholas 2001, 11).

Weitz states that "Project management defines the work processes and identifies phase and tasks required for a specific project. Project management determines the schedule, and monitors progress and status as well as refines the schedule" (1993, 67).

These definitions are just a sample of the many which are available throughout the project management literature. While each of these definitions is different, they carry a common theme of utilizing management processes, tools, techniques, and methods to deliver project requirements.

The definition of project management is further constrained in that delivery of project requirements must occur within an established schedule, budget, and the output or outputs must meet minimum defined quality requirements. Projects are also unique endeavors which occur once, rather than repetitive efforts such as requiring operation transformation processes. Project management is also responsible for the technical and overall system aspects of the operation.

Thus, the key project management and operation management and engineering management differentiators are the context which the management activities occur in. Project management occurs in a very constrained environment which is described as the (a) development of a unique product or service which has not been produced before, (b) involves a well-defined and constrained project schedule, (c) has an established budget, and (d) involves limited resources which include people, physical, and financial resources.

From this section, the reader should understand that the framework being considered generally defines the management context of the role. If the setting is a continuous and repetitive transformation processes, the role is one of OM. If the context is very technology-driven then the role is one of engineering management. Once the context becomes controlled as the development of a unique product or service, within a defined and constrained schedule, scope, and budget the role becomes project management. As such, what is the project manager's ultimate role and responsibility? We answer this question in the next section.

1.1.7 PROJECT MANAGERS' ROLE AND RESPONSIBILITIES

In the previous section, we discussed the definitions of and differences between OM, engineering management, and project management. From this discussion it is clear that these roles utilize and leverage similar processes, tools, techniques, methods, and human resources skill sets to achieve successful delivery of the various roles' requirements. To accomplish this, within a project environment, a project manager is utilized.

This section provides a general overview of what a project manager's role and responsibility is and how they are critical to the success of the project and development of a successful project culture. This discussion is general in nature as the project manager's role and responsibility is specific to the organization, the project team structure, the organization's norms and culture as well as the unique project being implemented. So what is the project manager's role and responsibilities?

To start with, project managers or project leaders are ones with the responsibility for ensuring that all planning is performed and ultimately the project performance (Ptagorsky 1998, 7). Taking this description a notch higher, the project manager is a strategic position within the company's framework ultimately responsible for successfully delivering the project outcomes and outputs within the agreed to timeline and budget. They are the project's chief executive—the project manager has the ultimate responsibility for ensuring that the project (a) delivers the agreed to output, (b) meets all defined quality requirements, (c) is delivered in the agreed time, (d) is performed within the approved budget, and (e) effectively utilizes all resources assigned to the project.

As the individual ultimately responsible for the project delivery, the project manager's key roles include (1) leading the project team, (2) managing the project planning and implementation processes, (3) monitoring project progress, (4) monitoring project outcomes and outputs, (5) establishing a project environment that fosters a culture of success, human resource development, and ethical conduct, (6) managing the project risk, as well as (7) leading communications which include all internal and external to the project team interchanges. The project manager is like a musical conductor who brings the various talents, skills, capabilities, tools, and techniques together into a highly effective and efficient team.

To achieve a highly effective and efficient project team, which meets all of the projects outcomes and outputs, the project manager's responsibilities are extensive. The long or short project path from project initiation to final delivery starts with the role of project scope development and acceptance criteria. As was discussed earlier in this chapter, before the detailed project planning can start, an SOW document must be developed and approved by the project stakeholder(s) as to what the project scope is and what the acceptance criteria are. The project scoping document also provides a high level project schedule, and approved budget. Obtaining this initiating document is a key project manager responsibility.

Once the project scope and acceptance criteria document is approved, the project manager is responsible for the development and delivery of the detailed project plan. In alignment with earlier discussions, the detailed project plan is all inclusive and comprises various subplans such as (1) detailed schedule and time management plan, (2) implementation plan, (3) procurement plan, (4) risk management plan, (5) quality plan, (6) communications plan, (7) human resource management plan, (8) stakeholder management plan, (9) budget management plan, and (10) change management plan. Ensuring that the detailed project plan is developed, approved, and accepted is the project manager's responsibility.

As the project progresses through the implementation, acceptance, and closeout phases the project manager is responsible for ensuring that the project plan is followed. Throughout the various project phases, the project manager's responsibilities include specific activities such as monitoring the project progress against the plan metrics, providing required internal and external communications, mentoring the project team, change control management, and fostering a successful project culture.

In summary, project managers are the chief executive officer of the project. Their role is to ensure the ultimate delivery of a successful project. As the project progresses though its life cycle, the project manager's responsibilities are broad and diverse from ensuring the project is on schedule, within budget, meeting all quality requirements and human resource needs to providing communications both within the project team and external to the project team. To be successful, project managers leverage their leadership skills, management capabilities, and various tools, techniques, and processes.

Project managers frequently fill various human resource roles as well. In this role, they acquire and release personnel as needed. They also motivate the project team personnel and help to resolve conflicts. Project managers often find themselves dealing with interpersonal sensitivities, lack of communication skills, and credibility issues.

Project managers also fill a financial accounting role. They are responsible for reviewing and approving purchase orders as well as managing project contracts and their associated financial implications. They also need to ensure that all incurred and actual costs are provided to the corporate financial group as well.

Project managers fill a broad spectrum of roles and responsibilities. Their leadership and management skills must be high to provide a structure where the project team can be successful.

1.1.8 NATIONAL, INTERNATIONAL, AND MULTINATIONAL PROJECTS

The 21st century is a global business environment. Across the world a constant exchange of products, goods, and services is continuously occurring. Cities, towns, and organizations are often multinational. It is an era where throughout the developed world communications is almost instantaneous and movement of goods across borders is a routine process. In this global environment, project managers may find themselves working within a local or national context, in an international context, or within a multinational context. Diversity is now the norm, not the exception.

As project managers move from working in a local or national context to an international or multinational context, they must adapt their tools, techniques, and processes to these different settings. That is, the project manager may find out that what are acceptable norms within the national context may not be acceptable within the international or multinational context. As an example, while being a workaholic who puts in more than 40 hours a week is the U.S.-accepted norm and project plans are implicitly built around this assumption, in many other parts of the world this is not the norm. France, as an example, has a 35 hour a week law that its citizens must follow. A U.S.-based project manager who moves to a France-based project would have to adapt his or her plans and personal work ethic views to the 35-hour work week law rather than the 40 plus hour a week, which is the U.S. norm.

Another area of difference is the acceptance or rejection of nepotism within the project team. While in the United States many organizations have explicit rules of where family members can work and whom they can and cannot report to, this is not the same throughout the world. In some countries and regions of the world, nepotism is not only accepted but is also encouraged (such as various Asian countries).

As the literature notes:

> There are a variety of barriers that exist when facilitating international projects; in fact, barriers exist within every culture, some obvious and others not so obvious. Making matters more difficult, the barriers change with each country and sometimes within regions of the same country. (Becker 2003, 52)

Understanding these variations is essential if a project manager is to be successful within a multinational project team environment.

While we expanded on the culture discussion further in later chapters for now the reader should realize that project teams are influenced and impacted by various cultures. These cultures include the national, local, organization, and project team specific structures. As will be discussed in more detail later, depending on the location and composition of the project team the various cultures may support or be in conflict with each other and subsequently provide an obstacle or support system for the project team's ability to be successful.

Understanding the various impacts, which may occur when international or multinational project teams are deployed, is a significant project manager issue. Being a project manager of international or multinational projects teams requires a unique skill set which can work within the different cultures.

1.1.9 PROJECT MANAGEMENT PROFESSIONAL ORGANIZATIONS

A quick Internet search on professional organizations reveals scores of different organizations within areas of management, technical, education, health care, engineering, human resources, science, and so forth. Virtually every professional field has one or more national and international professional organizations an interested person can join.

In general, professional organizations are, within the United States, nonprofit organizations which rely heavily on volunteers to keep the societies working smoothly and meeting the membership needs. These entities have many objectives such as being advocates for the members by directly working with the federal and state governments to educate the law makers as to the positive benefits of the society and to help shape any legislation which may positively or negatively impact the membership.

The professional organizations also foster the discipline's advancement by sponsoring conferences, workshops, educational opportunities, research opportunities, and providing magazines and academic level journals where authors and researchers can publish their work. Some professional organizations also develop industry-specific standards, recommended best practices, and reports as a means of reference and to provide a consistent approach to how efforts are conducted under similar situations.

Professional organizations also develop and maintain professional certification programs. These programs are typically founded on a specific body of knowledge that is directly related to the profession. To obtain certification involves not only passing an exam but also having documented a minimum level of employment within the field, and some organizations even require the certification candidate to be interviewed prior to being certified. The professional organization then ensures that the certificate holder maintains his or her skill level current by requiring a minimum level of continuing education and hands on experience every three to five years, depending on the certifying organization.

Another aspect of professional organizations is the establishment of a code of ethics that the members agree to follow. A code of ethics, according to Molander, is the "… principles of right and wrong conduct which guide the members of a … profession …." (1987, 619).

So professional organizations are advocates for their members, foster several means of professional advancement, and help to establish standards, guidelines, and codes of ethics the members ascribe to. It is a sign of a discipline's maturity and universal acceptance when it has a thriving professional organization.

Within the project management discipline, the two largest international professional organizations are the IPMA and the PMI. The following sections briefly discuss each of these organizations' history and mission statements.

The IPMA was started in 1964 when Pierre Koch of France, "…Dick Vullinghs from the Netherlands and Roland Gutsch from Germany [met]… to discuss the benefits of the Critical Path Method … " (IPMA 2005, 2). Over the next 40 plus years, IPMA has grown to include member associations from 55 different countries:

"IPMA is an international Federal, umbrella organization for national project management associations …. [who] represents … Member Associations on the global level" (IPMA 2005). Member Associations are nation-specific organizations that focus on their geographic locations to develop project management competencies within their unique cultures and in alignment with IPMA's overall guidance, decision-making processes, and governance structure.

The IPMA vision is to be the "… leading authority on competent project, programme and portfolio management (PPPM). Through our efforts, PM best practice is widely known and appropriately applied at all levels of public and private sector organizations" (IPMA 2005).

IPMA provides a four-level project management certification program where Level D is the lowest level of a Certified Project Management Associate, Level C is next as the Certified Project Manager, followed by Level B, Certified Senior Project Manager, and final Level A, Certified

Projects Director. IPMA's certification process requires that an individual (a) apply for a specific certification, (b) provide a list of applicable projects, (c) perform a self-assessment, (d) take either an optional or mandatory exam, (e) provide a 360-degree assessment, and (f) be interviewed. The process is to determine if the individual has the appropriate competence for the certification level applied for.

IPMA also provides a set of project management standards which include the IPMA Competence Baseline as well as the IPMA Competence Baseline for Consultants.

PMI was formed by five individuals in 1969. Since its formation, PMI's membership has grown to include about 265 geographic chapters and members in virtually every nation. The PMI organization structure includes a 15-member volunteer board of directors and an executive management group whose role is to guide day-to-day operations. From this central governance group local chapters are formed based on geographic location.

PMI provides certification in six different areas which include (1) Certified Associate in Project Management, (2) Project Management Professional, (3) Program Management Professional, (4) PMI Agile Certified Practitioner, (5) PMI Risk Management Professional, and (6) PMI Scheduling Professional. Each of these certifications requires that the applicants first determine if they have a combination of educational and applicable experience which meets the specific application requirements. In general, the higher the individual's secondary degree, the fewer the direct experience hours required.

Assuming that the individuals' self-assessment indicates that they meet the minimum requirements they then need to apply to take the exam as well as provide a resume which documents the application meets all required criteria. After PMI reviews the application and agrees that the candidate does meet the minimum requirements, he or she is provided authorization to sit for the certification exam. Once the exam is passed the applicant receives his or her certification.

To maintain certification, all certified individuals must maintain their current skills and demonstrate it through continuing education, volunteer activities, or work experience. By participating in various activities the individual earns professional development units (PDUs). Depending on the exact certification, the certified individual must obtain and provide evidence of a minimum number of PDUs on a three-year cycle.

PMI's volunteers have also developed a set of global standards which are available to any interested party. Some of these standards include the

PMBOK® Guide, OPM3®, The Standard for Program Management, and The Standard for Portfolio Management.

While IPMA and PMI are the two largest project management associations they are not the only ones. Other geographic localized project management organizations exist such as, but not limited to, The International Association of Project and Program Management, Association of Project Managers, Canadian Project Forum, and Performance Management Association of Canada.

As this short discussion highlights, the project management discipline has a range of different professional organizations which individuals may join. The various professional organizations provide similar roles and functions which include items such as being their member advocates, providing certifications as well as training, standards, and performance guidelines. The professional organizations also provide a forum for their members to become involved in.

SUMMARY

In this chapter, we covered a brief but full discussion on projects and project managers. We discussed how projects have been part of the human experience as far back as recorded history goes. These projects are a natural output of the human race's endeavors and desires to develop new and unique products, services, structures, facilities, and capabilities. Each of these efforts carried a level of uncertainty and various other constraints.

We also identified how project management transformed into a formal discipline with its own set of tools, techniques, and processes in mid-1950s. During that era, specialized project scheduling techniques were developed such as the PERT chart and critical path analysis. In the 1960s, professional organizations were formed, as did focused educational opportunities.

We also developed and provided definitions for projects, portfolios, and programs. After developing a baseline understanding of what defines a project, we moved into defining project management and what the project manager's role and responsibilities are. We then discussed the two leading international project management associations and how they contribute to the further advancement of the professional discipline.

The intent of this chapter is to provide the reader a firm foundation of what a project is as well as the role and responsibility of the project manager. This foundational knowledge is essential for the broader description and discussion of culture within the project environment.

CHAPTER 2

INTRODUCTION TO CULTURE: WHAT IS IT AND WHY IS CULTURE INVOLVED IN PROJECTS?

You can't walk alone. Many have given the illusion, but none have really walked alone. Man is not made that way. Each man is bedded in his people, their history, their culture, and their values.

—Abrahams (2013)

A people without the knowledge of their past history, origin and culture is like a tree without roots.

—Garvey (2013)

2.1   INTRODUCTION

The chapter's objective is to provide the reader sufficient information for a firm grounding from which we can build a deeper understanding of culture within the engineering project discipline as discussed in the next chapters.

2.2   HISTORICAL VIEW: CULTURE RESEARCH AND ITS ORIGINS

To start out, in this book when we refer to or discuss culture, we are not referencing the biologist's petri dish research, ethnicity, or social backgrounds, and we are not specifically discussing the *refined* taste that some refer to as in they appreciate the finer things in life such as grand art, fine foods, or the opera. In this book, we focus on the underlying values, beliefs, and shared philosophies that a group of interacting individuals shares. The cultural variables, features, attributes, or processes which we will be discussing are not directly measurable but have immense impact on how social groups interact and make decisions. As culture is not directly measurable, the course of developing, implementing, and communicating what culture is, is a difficult process with many different paths.

The many different paths of studying and researching culture are paved with efforts and views across a diverse background of academic disciplines, scholars, and philosophers, to name a few examples. The research in this area can be described as starting with some basic questions which are intended to identify what establishes the foundation of socially accepted behavior, within a group of associated individuals. Some of these questions include: Is the acceptable group behavior something we are born with or is it something we learn? That is, is the behavior genetically determined or is it nurture derived, that is, something that is learned? Other culture questions include:

- Are our accepted group interactions driven by a single trait, a few distinctive traits, or a broader set of attributes?
- Is our culture geographically based or historically based, and what causes it to change, or does it change?
- Are team, company, region, and national cultures similar, different, or the same?

Trying to derive definitive and quantifiable answers to these questions has engaged a spectrum of individuals and groups across a wide range of academic study areas, consultants, and practitioners. The first question we will address is where the word *culture* came from. To answer this question, we look at historical research.

The historical culture literature indicates that the first written use of the term culture, in respect to human behavior, occurred around 1871 by an English anthropologist Edward B. Taylor (O'Neil 2006). Specifically, "Culture...taken in its wide ethnographic sense, is that complex whole which includes knowledge, belief, art, morals, law, custom, and any other capabilities and habits acquired by man as a member of society" (Taylor 1871, 1). The literature is also clear that the study of human behavior based culture is firmly rooted in "... anthropology and is related to the individual's underlying values, beliefs, and shared philosophy..." (Henrie 2005, 2).

The origins and transformation of the meaning of culture to today's general consensus starts in the mid-15th century where

...culture [originally] referred to the tilling of land .... [It then changed, in] the figurative sense of "cultivation through education" [as was] … first attested c. 1500. [Culture's] Meaning [as] "the intellectual side of civilization" is from 1805; [and subsequently] that of "collective customs and achievement of a people" is from 1867 (*Online Etymology Dictionary* 2013).

As this short timeline shows, over a course of about 300 years the meaning and understanding of culture evolved from referencing the tilling of land to collective customs and achievements of the collective group. Throughout this evolutionary process, the concept that culture is a nurtured, versus natural, process remained constant. This point answers the question: Is our culture a result of our genetic makeup or our surrounding environment?

Historical research also indicates that culture was not a slowly evolving process but a rapidly changing social interaction event where change is sudden, not slow or evolutionary (Lehman 2010). It appears that the early human race transitioned from a set of individuals primarily trying to survive, to groups of people working and interacting as a social unit based on one of culture's artifacts, language. A culture artifact is something which transmits information about the group's culture, and Lehman's position is that human culture sprang up when we developed language (Lehman 2010).

From this early anthropological research, the study of culture expanded to include virtually every discipline, nation, state, organization, and team structure. As an example, active culture research is ongoing across an array of different academic and practitioner research disciplines such as sociology, history, management, and leadership studies (which include the organization and various team structures) as well as literary studies. Also, culture is being studied in virtually every nation across the globe, within numerous organizations, and in various team settings.

Along with the expansion of culture research to virtually all corners of our world and multiple disciplines the original culture description and how it applies within different social settings continued to evolve as well. As an example, based on the published literature, today's generally accepted understanding of culture did not occur until the 20th century. At that time, anthropologists assigned the concept of human aspects to the word. These aspects included our underlying values, beliefs, and shared philosophies. These human features are not based on nature, which means that they are not genetically transmitted, but are learned personal attributes which result in each individual's culture. As an example of this evolution process, it is seen that in the 1980s the ". . . ideas from mid-twentieth-century institutional and ethnographic studies of organization bore fruit . . . [with] the emergence of organizational culture frameworks that emphasized organizations as systems of meaning and symbols..." (Morrill 2008, 16), which are learned processes rather than individually explicit deoxyribonucleic acid (DNA) genetic codes.

The research is also clear that culture is a complex, socially interactive relationship. The complexity is derived from multiple factors, one of which is culture is not directly measurable and as such it must be inferred. That is, while you cannot directly measure someone's culture you can indirectly observe it at many different interaction points such as is found in national, organizational, social, corporate, or team structures. The different interaction points, in this concept, are how visible the culture phenomenon is to the observer. At one level the phenomenon is visible and demonstrated in a tangible form while at the other extreme culture is tacit and deeply buried within the very psyche (Schein 2004, 25). A factor which drives culture complexity is that part of it is that we just do things a specific way because that is how it is done. We cannot explicitly describe or define the reasons behind the action other than that is how things are.

As culture research expanded into virtually all social, organizational, national, and team areas of study, it was a natural progression that researchers would delve into project management culture.

Project management specific literature, in general, is a relatively young occurrence which can be linked to the establishment of various project management organizations, the formalization of project management as an acknowledged management knowledge area, and the establishment of project management as a discipline. As identified in Chapter 1, the two leading project management organizations were formed in the mid- to late-1960s. A literature review also identifies that "... there was a paucity of [project management] research in the 1960s and the 1970s. The research increased significantly in the 1980s and expanded in the 1990s" (Kloppenborg and Opfer 2002, 22). The path of project management culture research literature does not exactly follow the overall project management research and published literature trend but it and continues at a low level.

To describe the low level of published project management culture research, one can review and characterize the various project management literature sources. The outcome of this review confirms that during the early years, the 1960s and 1970s, there was a comparable paucity of culture research which matched the overall paucity of project management literature, in general. Specifically, a key finding of the first 40 years of published project management research identified that only 4 percent of it was associated with culture (Henrie 2005, 23). When one considers that there was a minimal amount of project management literature being produced during the identified time frame, this level of culture research can only be described as minimal, at best.

As project management literature publication rates started to increase significantly during the 1990s and 2000s, the percentage of project management culture published literature remained relatively constant and small. A review of project management literature published between 1993 and 2003, which included a "… combined total of 770 Project Management Journal® and International Project Management Journal abstracts and articles [determined that] approximately 4.5 and 8 percent, respectively, of the articles provide data or information on culture research" (Henrie 2005, 23). These publication rates are very comparable to the early research publication overall percentages. The percentages also demonstrate that project management team culture was not a significant portion of the research being conducted during the last two decades.

While project management culture has not received extensive treatment, it has been and continues to be a field of study. So, what has the research determined in regard to culture in projects?

Within the project management published culture literature, one of the earliest references to project management culture occurred in 1975 where it was identified that all members of the team bring with them their own culture and its associated beliefs, values, and attitudes. That is everyone arrives with what they see as acceptable in how to communicate, interact with others, their values, and how to make decisions (Morton 1975, 23).

This early project management view is in general alignment with the overall concept of culture. That is, team culture is based on the some core set of underlying values, beliefs, and shared philosophies. It is also in alignment with the organizational and team culture research which has identified how the various members develop a common set of assumptions, processes, and decision-making methods.

While project management culture research is relatively new and the breadth and depth of the research is limited, some common themes have evolved. One theme is that project management culture is predominately a Western-based product regardless of where in the world the project occurs. The project management journal articles by Henrie and Sousa-Poza (2005) as well as Wang and Liu (2007) are two such examples of the research in this area. This is attributed to the factors that the project management discipline originated in Western society and it has been exported around the world through Western-developed training classes.

Another common theme is the impact of training on changing a project team member's culture. "Previous studies have found that professional training could play a key role in reshaping people's work-related values/ beliefs..." (Wang and Liu 2007). These literature sources identify that those who receive formal project management training, determine that their project team culture becomes more aligned with Western culture than the local or national culture or the culture where the project is being performed. In other words, the research indicates that individuals modify their interpersonal and intergroup behavior for the project environment and based on the training they have received. Thus, the Western culture based training establishes a general Western-based culture in projects around the world.

As an example of how project-specific culture modification can occur, one can look at the area of interpersonal communication between the team member and the project manager. There is an extensive set of research studies on national-based culture superior to subordinate communications. This research identifies that on one end of the spectrum the subordinates are not only expected to but encouraged to challenge and provide active dialogue with their superiors. Yet, the other end of this communication range identifies that there are cultures where the subordinates are explicitly discouraged from any active dialogue. They have a "do as told" rather than "ask questions" culture.

In Western project management culture, the expectations and team culture are toward the active dialogue side. Cross-culture and international culture research has identified that project team members modify their project team communications toward an active dialogue environment. This communication culture modification appears to only apply to the project team environment and does not carry over into the broader organizational setting. A strong modifier to this occurs when the project team members have received formal project management training. As most project management training has a Western origin this is reflected in the training and the team communication culture tends to become more Western than if they have not. When the project team members are not interacting with the project team, their local communication culture once again becomes the predominate culture.

One other common project management culture research theme is the importance of the project manager in establishing the right cultural environment. As reported by Shore and Cross "... one of the roles taken by project managers is that of a "social architect" who must understand the role that behavioral variables play in project success" (2004, 56). This project management theme is consistent with organizational leadership culture research which identifies that "... culture begins with leaders who impose their own values and assumptions on a group" (Schein 2004, 2).

Development of a project success culture begins with the project manager and ends with the project team.

Recent project management literature also substantiates the overall position that within projects the primary key element to project success is the human. Regardless of what policies, processes, or procedures may exist it comes down to the team member. Yet, the literature also is clear that empirical studies of the human element and culture are lacking (Bredillet 2011, 2). The dichotomy that culture is critical to project success yet research is lacking in this area is partially driven by the fact that researching culture is difficult as researchers must rely on indirect measurements to infer the cultural attributes they are trying to study.

The challenge of understanding project management culture is high but the potential benefits, to the project environment, are high as well.

In summary, the history of culture research dates back at least to the early 1800s and has evolved or extended to virtually every type of social organization. The project management culture research is one of the newer culture research discipline areas with its origins in the mid-1960s. Across 50 years of project management culture research, some common themes have developed. These include items such as:

- Project success is dependent on the project team's common culture.
- The project team culture can be modified.
- The project manager is essential to the establishment and implementation of a successful project team culture.
- Project management culture is difficult to quantify and qualify as culture is not directly measurable.

We take on the challenge of defining what culture is and ways that we can infer the culture in the next section.

2.3 CULTURE: WHAT IS IT AND HOW DO WE DEFINE IT?

"Culture is one of the two or three most complicated words in the English language" (R. Williams 2013). To develop an understanding of what culture is, it makes sense to develop or adapt or in some other means provide a definition of what the word culture means as used in this book. To establish a definition, one generally looks to other literature to see if there is a common definition or usage that researchers, academics, and practitioners have adopted. By following this process one finds out that there is no single or universally accepted definition of culture within the literature base. As an example, one mid-1980s study identified over 300 culture definitions (Storti 1998). This was an almost twofold increase over a 1952 study which identified over 160 culture definitions (Bertalanffy 1969). While the growth of culture definitions is not as dramatic today, new definitions continue to be added every year.

With such a range of potential definitions and "Faced by [the] complex and still active history of the word [culture] it is easy to react by selecting one 'true' or 'proper' or 'scientific' sense and dismissing other sense[s] as loose or confused" (R. Williams 2013). Yet, while a single, universally accepted definition is not available, we must not sink into the morass of competing definitions but raise to the challenge of developing a definition which is applicable to this book for consistency and usability.

Rising to the challenge, we start with first looking at the origins of the word culture. As presented in the preceding section, culture's origins come from Latin *cultus* and French *colere*. *Cultus* is translated to mean care while *colere* means to till as in to till the ground (Berger 2000). So how does *cultus* and *colere* merge into culture, and again, what does culture, in relationship to project team accepted interactions, mean and where did it come from?

First, no one is born with culture DNA genetic coding. There is no genetic culture trait which establishes how we interact with society on the day we are born. Thus, we all start with a zero culture foundation and evolve from there. It also something that someone living alone on an isolated island would not have as culture is a shared concept. As a shared concept, developing a common culture requires interaction with others. If there is no interaction between individuals, then there is no ability to develop a culture or, more precisely, a shared culture.

Thus, as we are not born with an ingrained or genetic-based culture it is a learned concept. Learning one's culture can be related to growing something or as the French word *colere* means to till the ground or to grow crops. It is also closely related to the Latin term *cultüra* which relates to cultivation. Cultivation, in the social sense, generally refers to the process of refining, improving, or training a person in what is acceptable, refined, good, or desired as viewed from the perspective of the shared society as a whole. Therefore culture is something that occurs between people and is cultivated over time. As the social group interacts, over time, they define what is acceptable, not acceptable, how to make decisions, ethical and morale interactions, what acceptable communication is, and how to be successful within that environment. This culture assembly also passes the concepts on to those who join the social group. The new members learn the group's culture and correspondingly the group's culture undergoes continual change as well.

A key element in the understanding of culture is that it is not static. As social groups interact, new members join, others leave, what defines success changes, and how people interact with each other also changes. In the end, the group's culture changes as well. As the literature identifies, every organization, group, or team's culture, at any point in time, will change over time. They become somewhat unique in their interactions. As membership changes, the new group develops its own culture (Schein 2004, 274) and even when membership is static, the culture continues to change over time. An example of this can be demonstrated by how Western management culture has changed over the last 100 years. In the early 1900s, Henry Ford said that the customer could have his or her car painted any color they wanted as long as it was black. This authoritarian statement can be interpreted as indicating a culture of command and control where it did not matter what the client wanted—*I, as the person in charge, will make the final determination and that is your car will be black.* Counter that with the 21st century automobile manufacture culture: *What color do you want it to be*? Today's culture is grounded on the view that we are in business to satisfy the customer and the customer is right, so there has been a cultural change. The culture research also identifies culture as an interaction among the members of a group having individual's beliefs, values, and shared philosophies as they resolve issues, make decisions, and learn how to cooperate with each other.

In discussing beliefs and values, a blurring of definitions and intent occurs. Authors often use the terms interchangeably and in a general fashion, rather than in a prescriptive manner. Part of this blurring of terms and meanings occurs as the two words are closely aligned and form the basis of how we interact with others and what we accept as morally correct. Yet, there is a distinct difference between the two terms. "Beliefs are the convictions that we generally hold to be true, usually without actual proof or evidence.... and our values stem from those beliefs. Our values are things that we deem important and can include concepts like equality, honesty..." (DifferenceBetween.net 2013). In this vein, beliefs are described as the team's moral standards and norms. They are deeply ingrained and influence how we interact as we internally know them to be true. Values, on the other hand, are our ideas of what is important to us and establishes how we typically view important interactions.

Expanding on values a little further, when we see or hear the word *values* we immediately *know* what we mean by it and it takes a few minutes to verbally or in writing set forth what we *intuitively* know is the correct way to act. In general, our values define the ideas which are important to us such as what is acceptable, what is not acceptable, what it means to be honest or not. Predominately, we learn these values before the age of 10 and they are so ingrained in our way of thinking and acting that they become unconscious to us. There are also attributes that we find difficult to explain because they are our *values* and they form, for us, the basic aspects of right and wrong, what is acceptable and unacceptable, as well as the difference between honesty and dishonesty.

Organizations and teams also espouse a value system which establishes what is acceptable and unacceptable as well as what is honest and dishonest behavior within the group environment. One can view values as the rudder on the ship in that the rudder provides the direction for where the ship will sail. Values provide the individual the same guiding direction as they sail through the daily seas of opportunities and challenges (Deal and Kennedy 1982, 21).

Development of this common value system occurs in many different ways. For companies with a long history and well-established culture, new employees assimilate this culture through training and interaction. In new entrepreneurial organizations, the founder or group of founders establishes the company's culture based on their view of the world and how everyone should work together, make decisions, and the accepted risk level. Project teams are similar to the new enterprise in that the project manager is instrumental in establishing the project team culture. Values are the organization's foundation on which everything can be built (Deal and Kennedy 1982, 21). Within projects, values are the bedrock to any project team culture as well.

In this vein, the company philosophies provide the employees direction on how they are to do their work. These philosophies are built on aspects such as the founder or leader's internal values and philosophies, which set the organization apart from its competition and peers. They are the foundational values that distinguish, set apart, and endure (Margolis 2013).

Building on this company philosophy discussion, we can characterize a project team's philosophies as the stated values of how the team members will do their work. They include:

- That small set of values that are fundamental, distinguishing, and enduring to the organization. An example would be that the project team values honesty, integrity, and collaborative dealings both internal and external to the team.

- The manager encourages team input and discussions in the decision process but he or she takes full responsibility for the final decision.
- The team is distinctly unique in that the members are willing to embrace risk using sound analysis efforts and team involvement.
- The team is a collaborative and supporting structure where the team members work to enhance each other's skills and capabilities with the objective of enhancing the project team's working environment.

These values, beliefs, and philosophies directly affect how the team interacts internally and externally. They are real but difficult to measure as "… values can't be seen or touched" (Booz & Co 2013). They must be inferred or derived very similar to how other scientists and researchers determine if a black hole exists by measuring other physical elements. In the same vein, we deduce the existence of the company's or team's culture by indirect observations of physical events. Some of the manifested culture physical events or process includes artifacts, symbols, heroes, and rituals which the following sections discuss in more detail.

Artifacts, on the other hand, can be seen, they can be touched, and they can be clearly identified (Schein 2004, 25). Another way of stating this is that an artifact is any tangible thing which conveys information on how the team interacts. It can be tangible items such as the project-centric specific language used, how the office space is allocated, what posters are displayed around the work area, what posters or other items do the individual team members maintain on their desk and in their work areas, and whether there is a central location where the project team members naturally gravitate to for both informal and formal meetings. The key points are that the team artifacts (a) are tangible items, (b) reflect the team culture rather than the individual's culture, and (c) convey information on the team's interaction.

Exhibit 2.1 provides a simple comparison between two teams' observed artifacts. From this quick and high-level view, Team A's culture appears to be very formal with a strict command and control environment. On the other, hand Team B appears to be more open, values and encourages team collaboration, as well as focuses less on hierarchical stature. There is an apparent and observable difference in these artifacts between command and control, hierarchical, structure focused team, Team A, and the open, interactive, and collaborative Team B.

Symbols, although closely related to artifacts, are another tangible item which one can use to infer the project team's culture. Symbols are items such as lapel pins or pens which have the name inscribed on them. A

Exhibit 2.1. An artifact comparison.

Artifact	Team A	Team B
Office space	Sized according to team position	Equally sized
Wall posters	Predominately safety related	Open communications, team spirit, safety
Meeting location	Formal meeting rooms	Formal meeting rooms and a large common area which is predominately used for informal and teamwide discussions
Communications	Predominately formal and in writing	Ad hoc, impromptu, phone calls, formal when required

symbol may also be a unique emblem, name, logo, or quote which explicitly depicts the unique team. Again, the symbol is unique to the team and a physical representation of the team's culture.

How the team identifies, represents, and recognizes heroes is another tangible means of identifying the team's culture. One definition describes heroes as "… persons, past or present, real or fictitious, who possess characteristics that are highly prized in a culture. They serve as models for behavior" (Tamu 2013). Hero stories are used as examples of the extreme efforts one must make if one wants to become a hero as well as a framework of how things are done within the team environment.

The next culture visual indicators we will discuss are rituals and ceremonies. As with culture in general there are many different views and definitions of what rituals are. Predominately the term ritual is usually associated with religious ceremonies, rites of passage, and various oaths of allegiance or coronations. Yet, within the various cultures we discussed to this point they all contain their own set of what we term rituals.

To minimize confusion, for this book we describe rituals as those sets of precise or exactly defined sequences of events that the team members follow each time the situation arises. It carries the connotation that failing to precisely follow the ritual exactly as prescribed will result in a negative impact or a negative outcome will occur. An example of a ritual could be that a motto is repeated at the beginning or ending of a meeting. The motto is repeated at exactly the same point in the meeting using the exact wording and tone. Failing to follow the ritual as defined could result in confusion among the team members and a sense of loss as well.

Another ritual could be the prescribed actions a team always performs when a successful or less than successful test occurs. An example of one team ritual is where every member on the local football team must slap the school's name sign, which hangs at the entrance to the field, as they enter the field. Another ritual is that all meetings must start with a safety minute, followed by a cultural minute, and then the presentation of the meeting agenda. This ritual reinforces the team's commitment to safety and diversity, which rank higher within the team's culture than the actual project they are working on. The team also feels that failing to follow these prescribed processes, exactly, will result in the team losing focus on what is critical to them and negative outcomes may occur.

In this book, rituals are the given processes which "… focus attention, establish significance, and achieve a beneficial result" (King 1997). They achieve an almost religious position within the team and the exact process must be followed every time. Ceremonies are closely aligned with rituals and are another visible indicator of the team's culture yet lack the religious fever that a ritual transmits.

Ceremonies can be formal or informal activities the team follows on special occasions. They are different from rituals in that while they are a prescribed process and occur every time the trigger event occurs, the participants can and do vary the words and processes rather than follow a precise ritualistic application. An example of a ceremony could be how a new member is welcomed into the team. This ceremony could be as simple as calling everyone together and introducing them to the new person or it could be an elaborate affair as in holding a welcome-to-the-team afternoon party. Another ceremony could be where the project manager takes everyone out to dinner when critical project milestones are passed. The key attributes to the ceremony are that (a) it occurs every time the trigger event happens; (b) it follows a standard format such as a meeting, a party, or a presentation; (c) the participants can and do vary the actual process as a means of tailoring it for that trigger event; and (d) it reinforces the team's culture.

To recap what we have covered to this point, culture exists everywhere. By our association and interaction with others in our society, within our organization, and within our project team we assimilate and help build what is acceptable and not acceptable within these settings as well as how decisions are made. Culture is dynamic. That is, suddenly or over a longer time frame cultures are able to, and do, change. The change can be driven by significant events which occurred within the group or external to the group or it may be driven by the maturing of the group as a whole or changes in the group's membership composition. An essential project manager role is to establish and foster the team culture which solidifies how the project team will interact, make decisions, as well how the team will handle positive and negative events.

Establishing, identifying, and changing the team's culture is a difficult process as culture is not directly measurable. The engineer or project manager may *know* what the team values and beliefs, that is the culture, must be but he or she has to rely on the observable, not unobservable, manifestations which indicate what the culture is and utilize this knowledge to affect the culture transformation.

As noted in the preceding paragraphs, the observable culture manifestations include items like the team's philosophy statements: its heroes, symbols, rituals, and ceremonies. Each of these is an essential and critical aspect of the project team's overall culture and provides a means of indirectly measuring or identifying the team's culture as well. As such, the project team's philosophy provides the guiding *light* or framework on how the team will work together. The team's heroes identify what team member characteristics are most desired. Symbols and artifacts provide visual strengthening of the culture, while rituals and ceremonies are processes which maintain the team's culture as tangible and real.

2.4   CULTURAL THEORIES AND DEFINITION

Based on the foregoing, the reader should have a better understanding of what culture is but at this time we have not provided a definitive culture definition that will be used throughout this book. To achieve this, we look at the three major culture areas of research which consist of (a) national culture, (b) organizational culture, and (c) team culture to obtain a broader understanding of what current literature and research has to provide. This discussion starts at the macrolevel, which is at the national level, and progressively moves toward the microlevel, which we define as the team. This progression does not imply, indicate, or explicitly show that each level directly influences the more consolidated group of interacting individuals or that the smaller groups directly alter or influence the broader entities. How these various groups' cultures may or may not interact is a topic of discussion in later chapters. For now, we apply the national to team progressive discussion as a structural approach, only.

Yet, before delving into the details of the various groups' culture research, theories, and definitions we need to take some time to explicitly define two critical words which are often used interchangeably and sometimes incorrectly. Specifically, the two words which are often used interchangeably yet have distinctly different meanings are theory and hypothesis.

A theory is a principle which is universally accepted as something which clearly explains the objective of analysis. As an example, gravity can be described as a theory as many different observations and scientific tests continue to produce the same answer. From this one can proceed forward and apply the law of gravity to other analysis efforts and have it accepted as reasonable and valid.

A hypothesis, on the other hand is a testable prediction which you are testing. As an example, it is a hypothesis that at the end of this chapter your knowledge of culture will be improved. This hypothesis can be proved or disproved based on a series of tests. Yet, the results of these tests will not establish this hypothesis as a theory since it may not be universally proven across many different settings.

- "A theory predicts events in general terms, while a hypothesis makes a specific prediction about a specified set of circumstances."
- "A theory has been extensively tested and is generally accepted, while a hypothesis is a speculative guess that has yet to be tested" (Cherry 2013) or more precisely still has room for doubt even when tested.

While one may tend to view some of the following material more as a set of tested hypotheses than a set of theories we will rely upon how the literature generally tends to describe the discussion. As such, the following discussion is set in the tone of theories, rather than tested hypotheses. It is not this book's intent to enter into the discussion that these are theories or tested hypotheses but relay the information as found within the literature.

With this clarification in place, we now take a look at the national cultural research of Dr. Geert Hofstede as well as the combined work by Dr. Fons Trompenaars and Dr. Charles Hampden-Turner. At the organizational culture level, we review the research of Dr. Edgar H. Schein as well as the collaborative efforts of Terrence E. Deal and Allan A. Kennedy. For the project team culture research we will review the general trend of research, in this area as, as is explained in more detail later, there is no single researcher or group of researchers who have emerged as leaders in this area of study.

Further, the discussion on national culture, organizational culture, and project team culture is held within these definitive boundaries as the research identifies that each is unique and may not be interrelated. By way of explanation, national culture is assimilated by the individual before the age of 10 as an interaction with family, friends, school, religious settings, and social interactions.

Organizational culture occurs as the individual interacts within the work environment or other structured environments—for example, boy or girl scouts or other social environments. Within the workplace, the individual develops or adopts the company's norms, values, and beliefs. Research has identified that national culture and organizational culture can be different in six dimensions (Minkov 2011, 14).

Team culture can also be different from the work organization's culture as well as the national culture. Team culture, as with work organizational culture, is acquired when the individual works within the project team environment. It is a rapidly forming culture which is consistent with the unique and short-term context within which most projects exist. This culture is temporary and may only exist within that project's time frame as the project will end and either a new project team will be formed to work on the next project, which will result in a different culture, or the individual will return to the broader organization and its unique culture will prevail.

As a note to the reader, the following discussion focuses on a small group of cultural researchers. There are many other leading researchers whom, for reasons of brevity, this book does not provide an introduction to.

2.4.1 DR. GEERT HOFSTEDE: SOFTWARE OF THE MIND

At the national culture research level, a leader in this area is Geert Hofstede. Dr. Hofstede's culture research has its origins in his organizational work as an employee of the International Business Machine (IBM) Corporation. Within IBM, Hofstede worked as the manager of personnel research and as a management trainer. One of Dr. Hofstede's IBM roles was to introduce and oversee the application of employee opinion surveys in an effort to understand how employees worked together, their behavior, and collaboration processes. In this role Dr. Hofstede was also required to travel to more than 70 IBM subsidiaries worldwide to perform employee interviews. These interviews provided firsthand information and accounts of how IBM employees interacted and collaborated in these different settings. As reported in various literature sources, the IBM research generated a database of over 100,000 completed questionnaires, from 72 different nations. From the analysis of this extensive database Dr. Hofstede created his cultural dimensions theory.

Dr. Hofstede's original cultural dimensions theory moved away from the common cross-cultural research view that culture was a single attribute or variable (Minkov 2011, 10) to that culture can be analyzed and characterized based on a set of dimensions rather than a single variable. This change in approach, from this author's view, alters the discussion of culture from a mental image of a flat plain to a view of an entity which has breadth, depth, and length. It represents a volume which is more inclusive of a description than an area-only view. Dr. Hofstede's original cultural dimensions theory is based on four dimensions:

1. Power distance
2. Individualism–collectivism
3. Masculinity–femininity
4. Uncertainty avoidance

It is important to note that the dimensions as distinct entities are not physical, tangible things but are items which are inferred or concluded from indirect indicators such as verbal statements, observations of various behaviors, or how respondents answered various survey questions (Hofstede 2013). For each of these dimensions, Dr. Hofstede establishes an index rating where each nation can be compared to the others. The index rating was originally based on a 0 to 100 factor analysis statistical result score. Subsequent to the early efforts the upper range has been extended to 120 based on further research.

Yet, what information does each of these dimensions provide? To answer this question, the following provides a short discussion of each dimension. Each discussion presents a set of referenced countries and their index rating for illustration purposes and to highlight how nations are compared on each scale.

The power distance dimension defines how the nation handles social status and authority power. Across the globe people live within nations where social power exists at various levels and with varying degrees of inequality. The power distance index provides a means to compare different nations' accepted means of how they interact within this inequality. Nations with high power distance index values, such as Russia with a 93 and Philippines with a 94, tend to be societies which place everyone in a hierarchical order and no further justification is required as that is how the culture is. This is dramatically different than low power distance index societies such as Denmark with an 18 and Israel with a 13.

In these societies, hierarchical order is minimized and social equality is more the norm.

Hofstede's second dimension is individualism versus collectivism. This dimension represents how individuals and society interact. A high individualism society exists where individuals are expected to take care of themselves and their immediate families and not to rely on the broader society as a group or whole to take care of them. The United States with an index of 91 and Australia with 90 are examples of high individualism nations. Collectivism nations, on the other hand, generally view the broader entity as a collective whole, in the light that it will take care of the individual in exchange for continual loyalty to the broader society. Russia is an example of a collectivism society with an index of 39 as is China with 20. One way of viewing this dimension is to think whether I see my interaction within the group in the view of "I a single entity" or as "we the group."

The third dimension of masculinity–femininity provides a national comparison index as to the nation's preference toward a competitive versus cooperative context. Nations with a high masculinity index, such as Austria with a 79 and Japan with a 95, value competition, work and school achievement, and success within their various endeavors. The masculine index identifies that, generally, it is a cultural driver to be the best possible versus reducing expectations to achieve a collaborative result. A clear example of Japan's culture is the extreme competition to excel in school and be accepted to top tier colleges. Thus, individuals and organizations are highly competitive and winning is imperative. These nations tend to associate success with material reward and heroism. On the other end of this cultural dimension, are nations that exhibit greater cooperative cultures, for example, Denmark with an index of 16 and Finland with 26. These nations strive for a cooperative, versus competitive, lifestyle. Consensus is prized and rewarded rather than the need to be competitive. They view working together to achieve a common goal of greater importance than competing with each other.

Hofstede's fourth cultural dimension is uncertainty avoidance. As the dimension name indicates, this dimension provides a means of comparing nations on how comfortable they are with vagueness and ambiguity. Countries such as South Korea with an 85 value and France with an index of 86, indicate that they are not comfortable with an uncertain future. As such, these nations tend to have very rigid codes, extensive rules, and a tendency to avoid or resist innovation. Security is very important within high uncertainty avoidance index nations. On the other end of the uncertainty dimension scale, there are nations like Vietnam with a 30 and Sweden with a 29 indicating they are very comfortable without sets of rigid codes. They also tend to believe that the number and extent of social rules should be minimal and definitely no more than that is required. They also view innovation as a positive attribute and are very comfortable with vague or ambiguous future states.

The key to these initial dimensions is that they provide a means to compare nations within a common framework. They are not intended for and do not work as a means to compare individuals within or across organizations and nations.

Hofstede's original dimension theory remained unchanged until 1991. In 1991, a fifth dimension was added: long-term orientation.

The fifth dimension, long-term orientation, was developed in a collaborative effort with Michael Bond (Minkov 2011, 12). This collaborative effort came about when Michael Bond developed the Chinese Value Survey which was then applied within 23 countries. While results of this survey provided further support for the original dimensions, it identified a new dimension, how the nations view the past and present. That is, does the nation focus its efforts more on the past or more on the future?

Countries, like the United Kingdom with an index of 25 and Spain with an index of 19, have a strong view of tradition and the need for quick results. They generally do not look toward the long term and save for the future. Conversely, nations such as China with an index of 118 (here we see the expanded scale which was described earlier in this chapter) and Taiwan with 87 have a long-term view of things. They see a need to save for the future and show a strong tendency toward persistence and perseverance rather than a fast fix.

An essential key to Hofstede's cultural dimensions theory is that the results are only applicable for comparing nations, not for understanding the individual or organizations within that nation. While, at the macrolevel, national cultures can be compared, using these dimensions, the analysis does not translate into the ability to understand a specific organization's culture or the individual's culture based on these same dimensions and index values.

Based on the five dimensions, how does Dr. Hofstede define culture? At the national level, Dr. Hofstede defines culture as *mental programming of the mind*. There are several key attributes to this definition such as it clearly implies that culture, as described earlier, is a learned process. As one programs a computer to perform specific things, we as individuals are *programmed* to interact with our major social group in acceptable manner. It also implies that this *programming* occurs early in your youth. As with computers, when they are first assembled, they have the potential to do many things but they are incapable of doing anything until they are programmed. The same is of newborn babies; they have great potential capabilities but until they receive programming these capabilities are not achieved.

Programming of the mind also infers that making changes is possible but may be hard to achieve. One has only to discuss the challenges of modifying an existing, complex, program with a programmer to understand the challenges and difficulties of this task. The individual's core culture is thus formed early in life. As such, many of the individual's core values are tacit in nature to a point where the person is unable to provide a detailed explanation why their values, norms, and beliefs are what they are. In fact, they are not aware of the ways that these early year learned culture attributes affect their behavior. As such, these foundational set of ethics, norms, and beliefs are very difficult to change (Henrie 2005, 128).

2.4.2 FONS TROMPENAARS AND CHARLES HAMPDEN-TURNER

Fons Trompenaars and Charles Hampden-Turner form an internationally known collaborative team of cultural researchers who developed a cultural model based on the following seven dimensions:

1. Universalism versus particularism
2. Communitarianism versus individualism
3. Neutral versus emotional
4. Diffuse versus specific cultures
5. Achievement versus ascription
6. Human–time relationship
7. Human–nature relationship

Dimensions one through five define the relationships among people in a society. The sixth dimension focuses on society's attitude toward time while the seventh looks at society's view on the environment. The seven dimensions' distinctions are based on Trompenaars' and Hampden-Turner's view that culture is how the group derives solutions to problems or dilemmas. In general, these problems or dilemmas are either associated with how one interacts with others, how things change over time, and the environment (Trompenaars and Hampden-Turner 1998, 8).

The seven dimensions model was developed through the analysis of data obtained in over 30,000 survey results which encompassed organizations from 50 different countries. The Trompenaars and Hampden-Turner data sources were from a variety of organizations and their research view is to explain cultural diversity within the business structure.

The dimension of universalism versus particularism takes into consideration how relationships are viewed. That is, does society view the need to follow the rules as more important than personal relationships? If this is the case, then that culture is described as a universalism culture. A society with a particularism dimension has the opposite view in that they feel that their personal relationships far outweigh society's rules. Family bonds, friendship, and long-standing work relationships mean more to them than a set of rules

Communitarianism versus individualism considers if society is more a *we* structure or an *I* structure interaction. What this says is that the communitarianism society is a *we*, or group-based, society. In this society the formation of groups, teams, and collaborative working groups are the norm. The group focuses on what they view is best for the group rather than the individual. Societies in Asia exhibit a strong communitarianism social structure.

Individualism, on the other hand, involves societies where the view is predominately from the *I* perspective. That is, society views the needs of the individual above the larger group. Success is defined as what the individual does and the personal responsibility he or she takes for his or her actions. In an individualism society, the stature of the individual can be viewed by the number of assistants they have.

The neutral versus emotional or affective dimension involves how we display—or not display—our emotions and feelings. Neutral societies are defined as those that do not openly show their emotions. The British *stiff upper lip* phrase is an excellent metaphor for this dimension. To display a *stiff upper lip*, you will remain resolute and unemotional in the face of adversity, or even tragedy. As such, you will be neutral and will not show your emotions. Emotional or, as they are sometimes referred to, affective societies are very open with their emotions. People in an emotional society have no issues with laughing out loud or crying in a public setting. They clearly reveal their emotions for all to see. Italy has been described as an emotional society. When two Italians meet there is a great showing of emotions and long conversations ensue. These conversations include a lot of facial expressions, body movements, and grand exclamations.

A diffuse versus specific culture takes into consideration how and to what level we interact with others. Countries with a diffuse culture view interaction between people as reaching beyond the current situation. As an example, the supervisor–worker workplace relationship extends beyond the workplace and permeates into their personal lives. In a diffuse culture, the boss has no issue with being invited to the worker's house for a party or special occasion. These societies tend to have lower workplace turnover as a job change impacts the personal level and many social and interpersonal connections and relationships.

Specific cultures view the interaction of people differently from a diffuse culture. That is, workplace relationships are contained within the workplace and family interactions remain within the family, not the broader society. In the case of a worker inviting a supervisor to dinner, the specific culture supervisor would generally decline the invitation as he or she would not see this as acceptable behavior. Specific-based cultures also tend to be more precise and direct in their interpersonal interactions. They are viewed as more transparent in their dealings.

The achievement versus ascription cultural dimension considers how social status is assigned. Achievement-based societies assign social status based on what the individual has accomplished. Individuals who have contributed or achieved more are assigned a higher social status than those who are viewed as not significantly contributing or are underachievers with a lower social status.

Ascription-based societies assign social status not based on what you have recently achieved but items such as where you went to school, what family you were born into, kinship, whether you are a male or a female, and your age. In an ascription society, people will find it difficult or impossible to advance in social stature by personal achievement if they went to the wrong school or were born into the wrong family or even if they are of the *wrong* gender.

Human–Time relationship cultural dimension describes how we view past, current, and future time. That is, we may see time as either a sequential series of events or as an interrelated, synchronic, process where things happen in parallel. Societies with a strong historical time view see the present and future as a repeat of the past. Those societies which have a present or current time assessment of past events believe that they lack a strong shaping force. The future as well does not hold a strong modifying force as they view the future as something yet to come and it lacks the ability to focus efforts beyond the current day. Future-focused societies consider the past as very important but all efforts look at the future.

This cultural dimension also takes into consideration how society structures time as either sequential or synchronized. Sequential-oriented societies view time as a series of passing events where one leads to the next. The sequential-oriented societies take time commitments very seriously and ensure they have planned things. Planning and order are required attributes in this society.

Synchronized societies, on the other hand, view time as interrelated. People in these societies view time as relative where being on time for appointments is not a strong characteristic. In this society, it is not seen as an essential effort to set specific time commitments and then sticking with them. Changing plans and altering time commitments are seen as acceptable and to be expected.

The human–nature relationship, or as it is also referred to as the internal–external dimension, identifies how society views nature and whether it should be controlled or not. Internal-based societies take the position that they can dominate and control nature. This society focus is on the person over nature.

External-focused societies take the opposite view in that nature controls the situation versus being controlled. These societies position their actions toward others and the environment rather than internal to themselves. The literature identifies one way to view this dimension as through the lens of *locus of control*. Locus of control is the concept of what results in the positive or negative aspects within their lives. Internal locus of controlled societies believes that it is in control while external-focused societies take the position that the external environment is in control. Internal-based societies believe that when they are successful it was through their specific actions. External-based societies view success as the result of external influences.

Within culture literature it is reported that Trompenaars began with earlier efforts by Parsons as well as Klukhon and Strodtbeck. One can say that these researchers viewed culture as how people solve human relationship, time, and environment problems. Based on this earlier work and analyzing an extensive set of surveys and data from training classes, Trompenaars and Hampden-Turner developed this seven-dimension cultural model. These seven dimensions are grounded in the concepts of relationship with others, the relationship with time, and the relationship with the environment. This approach provides the cultural researcher, student, and interested individuals one approach to understanding national culture. As such, one must always keep in mind that culture must be understood within the unique context in which it occurs and as a whole. It cannot be viewed as an island in and of itself or a set of individual attributes which have no interaction with the others (Hampden-Turner and Trompenaars 1997, 152).

We now move our culture research review from leading national culture researchers to the organizational culture researchers. Organizational culture research is more recent than national culture research. That is, organization culture research started after researchers began to investigate the impacts of national culture. These post national culture research origin efforts are only a few decades old with roots in the researchers' question of why are U.S. companies failing to perform as well in other societies as the local companies especially when national culture alone is insufficient to explain the difference (Schein 1990, 109). From this question came the attempts to explain the poor performance based on culture differences. To understand the various causes fostered the need to look at the organization's culture as national culture classifications, attributes, and understandings were inadequate to answer the core question.

This emerging area of study was identified in 1978 when Dr. Schein provided some of the first significant research in this area. From this emerging field of study, Dr. Schein's efforts have provided key insights into an organizational-specific cultural model.

2.4.3 EDGAR H. SCHEIN

As noted earlier, Edgar H. Schein is one of the leading organizational cultural researchers. His organizational culture model is based on the attributes of (a) artifacts, (b) assumptions, and (c) espoused values. These are higher cultural levels which are more visible to the researcher versus cultural concepts of values, norms, and beliefs. Each of these higher level attributes is described as a visible level which is easier to observe by researchers, project managers, or organizational managers. It is important to note that Schein's cultural model layers do not negate the concepts of values, norms, and beliefs, but that in Dr. Schein's model, these concepts are located between the visible layers and lower down in the overall structure which renders them harder to observe.

Dr. Schein's model begins with artifacts which are the things that can be physically sensed, observed, or touched. These are those things which we see as unique, different, or the same when we visit another culture (Schein 2004, 25). That is, artifacts can be the group's unique language or the specific way the team creates a project plan where the created project plan is an observable artifact. The artifact can also be how the project team structures its project life cycle's specific phases. How the life cycle is structured and how the phases are executed are can be viewed as the project's artifacts. How the organization establishes office space and the building it is housed in can be cultural artifacts as well. The key is that these are the things that one will see, feel, or hear when one comes into contact with the organization.

Dr. Schein also presents culture as a set of basic underlying assumptions which are those things that are taken for granted and are consistently applied within the organization itself (Schein 2004, 31). What this is referring to is that over time and through the process of solving similar problems the organization develops a consistent solution. The solution method becomes so ingrained that it becomes almost unconceivable that someone would solve it any differently.

Dr. Schein's next culture level is that of espoused beliefs and values. An interesting aspect of this cultural level is acknowledgment that culture has an origin or beginning. That is, it started somewhere with some interaction, idea, belief, value, or view of how interactions should occur to create a desired outcome. As the organization faces a new challenge, decision, or risk various individual cultures will be in play to derive an acceptable solution. Over time a group acceptable solution is developed. This establishes what the group views as what is the right or wrong way as well as what is acceptable or not (Schein 2004, 28). Within the published literature the reader will find this cultural level referred to as espoused beliefs, espoused beliefs and values, and espoused values.

Espoused values are developed through a process that the organization follows which transfers an original belief or value into the broader group's cultural espoused beliefs and values. The process starts with an issue which must be resolved. As an example; an entrepreneur, who is starting a new firm, believes that safety is paramount and that no work will occur if it is not safe to do so. Further, the entrepreneur's values say that it is better to stop all work than to have someone hurt. At the foundation level, this issue can be seen as the conflict between getting the work done and safety.

With the issue identified and the entrepreneur's espoused beliefs and values established with the group, the next step is implementation of the original beliefs and values. Thus, when a work event occurs, that appears to be unsafe, the responsible person takes action to stop work. This is implementation of the espoused beliefs and values. At this point there are two general paths which may occur. One path is that the person who took action to stop the work is acknowledged and rewarded in some manner. The other path is counter to this—the person who stopped the work receives negative feedback that his or her action was not proper.

In the case where the employee receives positive feedback, the espoused belief and values are reinforced. The group sees that when this situation occurs it is an acceptable and encouraged action to stop work.

This reinforcement feeds on itself and each time a similar situation occurs and similar actions are taken, with subsequent positive feedback, the espoused belief and value becomes stronger. Ultimately, the culture of safety transitions to the next layer of being a basic underlying assumption of how the company performs its work. From the group members' view this is just how they do work and no further explanation is required.

Conversely, if the alternative path is followed and negative feedback is received the group will reject the initial espoused belief and value. The group may view management's position as one of saying one thing and doing another.

Exhibit 2.2 provides a view of this feedback cycle. As discussed, it all starts at point A where the original espoused belief and value is provided to the group. At point B, the conflict between the espoused belief and value and the actual work event occurs with subsequent action taken. At point C, the work receives either positive or negative feedback. As shown, if the feedback is positive it will reinforce the espoused belief and value, point F, which continues the cycle until it becomes a basic company assumption. Conversely, if the feedback is negative the group will view the espoused belief and value as a process of management saying one thing and acting differently. In this case, the espoused belief and value will probably be rejected and never transcend to becoming a basic group assumption.

Summing up our discussion on Dr. Schein, he defines organizational culture within the attribute of basic assumptions. These basic assumptions are the group's learned set of behaviors which arose out of the group's efforts to solve problems. The basic assumptions now form what is acceptable behavior, what is right or wrong, and what is acceptable in interacting with the environment (Schein 2004, 12). He structures his discussion and cultural model around the three major cultural levels of (1) artifacts, (2) espoused beliefs and values, and (3) basic underlying

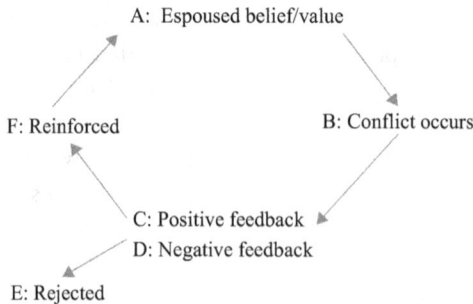

Exhibit 2.2. Espoused belief and value feedback cycle.

assumptions. As in other cultural models, artifacts are those things which are created by the culture and are observable. Espoused beliefs and values are those items which have no means to be empirically tested but must be accepted more on shared experience and group acceptance. Basic assumptions are the result of the group's empirical testing and validation of original beliefs and values. The empirical testing has validated the beliefs and values to a point where they become basic assumptions of how interactions will always occur. It is further identified that the organization's culture has an origin which is assimilated within the group over time. This assimilation or evolutionary process continues for the life of the organization. New challenges, issues, or conflicts create the opportunity to revise, add to, or change what the organization agrees is the correct way to react and act.

Dr. Schein's culture view is very consistent with other researchers in that culture is a very powerful abstraction. How the organization faces and solves problems and dilemmas is grounded in its learned culture (Schein 2004, 3).

In the following section, we expand the organizational culture research discussion to include the collaborative work of Terrence E. Deal and Allan A. Kennedy.

2.4.4   TERRENCE E. DEAL AND ALLAN A. KENNEDY

Terrence E. Deal and Allan A. Kennedy's collaborative efforts predate the early 1980s with the release of *Corporate Culture: The Rites and Ritual of Corporate Life*. Reviews of this book establish it as one of the first if not the first book which placed the discussion of corporate culture on organizations' radar screen. At the time this book was released Terrence Deal was a professor at Peabody College at Vanderbilt University. Allan Kennedy was the president of Selkirk Associates, Inc.

Deal and Kennedy present organizational culture within the context of four cultures which include:

1. Work-hard, play-hard
2. Tough-guy macho
3. Process
4. Bet-the-company

The work-hard, play-hard culture is characterized as occurring in organizations where the work is fast paced. In this culture, action is the key word and the employees tend to have higher stress over the amount of work they have rather than uncertainty of their actions.

The lack of uncertainty stress is associated with the tendency for low consequences to the organization due to failures (a low risk of failure). In general, each of the individual actions taken has minimal major consequence risks. While an error may occur it will probably not be catastrophic in nature.

Part of this low-risk nature can be associated with the third attribute of this culture and that is, it includes very fast feedback on the actions taken. A sales force organization has been characterized as a work-hard, play-hard culture group. This group generally knows that the sale is made or lost in a very short time frame rather than waiting around for years to know if the decision was correct or not.

The tough-guy or macho culture is significantly different than the work-hard, play-hard culture company. In the tough-guy culture there is a high-risk environment. Decisions and actions required are made by those who are willing to take major risks. They see this environment as supporting, taking big chances with the expectations of big rewards. Conversely, along with the potential for big rewards is the chance of major failures as well.

In the tough-guy culture, feedback is also rapid. Thus, as high-risk efforts are pursued success is quick to happen as well as failure. Success is rewarded and heroes emerge. People who thrive in this culture have a very near term focus rather than future state view. They know that major success or major failure can occur on the next decision or action taken. They are willing to roll the dice and see if they win big or lose big. The gamble is worth it to them.

In the process culture, things are very different in that the work pace is more relaxed. Metaphors such as plodding, secure work environment, benign conditions, and a comfortable work environment have been used to describe this work culture. There is little risk associated with the organizational activities and feedback is slow if ever received. Bureaucratic organizational structures are generally characterized within this culture setting.

The bet-the-company culture, as the name implies, is a high-risk organization. In this structure, decisions can have significant or catastrophic results. A wrong decision could result in the company's demise.

The bet-the-company culture differs from the tough-guy macho culture in the area of feedback. While the macho culture has a rapid feedback loop, the bet-the-company culture feedback is slow to occur. In the bet-the-company culture, years may pass between when the decision to act was made and results of that decision are made known. To succeed in this culture, one must be capable of extended stress periods as the uncertainty time frame is long. You must also be able to withstand the stress that your decision may put the company out of business or a willingness to quickly change jobs before the results are manifested.

Deal and Kennedy's cultural model is partially based on the concept that every business has a unique environment. As an example, there are companies whose primary environment is sales. A real estate firm would be one example of a sales-dominated organization. There are other companies whose primary business is research and development. Most pharmaceutical firms could fall within this general business environment as they engage in long-term research efforts where the final outcome may not be known for many years. Then there are organizations that have been and will continue to methodically plod along doing exactly what they have been doing for years. These firms have a low-risk and low-stress environment. Then there are organizations where people thrive on high-risk, fast response environments where almost every decision can result in a major success or stupendous failure. Understanding the variation of the business work environments is critical to understanding the company's culture as "Each company faces a different reality in the marketplace depending on its products, competitors, customers, technologies, government influences, and so on" (Deal and Kennedy 1982, 13).

Another key element is that values are the very heart of the corporation's culture. These are the basic concepts and beliefs within the organization which structure how things are performed, problems resolved, and individual interactions occur (Deal and Kennedy 1982, 14). This element is in alignment with earlier discussions in this chapter and is maintained throughout the book. This aspect is also in alignment with other literature sources which present how "Values are multifaceted standards that guide conduct in a variety of ways" (Rokeach 1973, 13). As the heart of the corporate culture, values guide us by:

1. Leading us to take particular positions.
2. Guiding presentation of the self to others.
3. Providing means to evaluate and judge.
4. Provide a standard to ascertain whether we are as moral and as competent as others.
5. Providing a means to persuade and influence others.
6. A means to rationalize in psychoanalytic sense, beliefs, attitudes, and actions that would otherwise be personally or socially unacceptable (Rokeach 1973, 13).

Culture research has also determined that companies with a strong culture will have a strong set of values. The value foundation identifies the firm's foundation for achieving success, establishing its vision, defines how one will interact within the organizational structure, as well as a common direction for all to follow. Success and failure are grounded in the organization's values (Rokeach 1973, 21).

With this brief review of organizational culture researchers, we now focus on project team culture research.

2.5 PROJECT TEAM CULTURE RESEARCH

The previous sections provided the reader with a short review of leading national and organizational cultural researchers and their associated models. While there are debates on the validity and applicability of these researchers' work, as well as that of other national and organizational researchers, there is a rich research focus within these areas. The literature is also consistent that national culture research predates organizational culture research which predates project team culture research.

In this vein, we find project team culture research is a very recent addition to or expansion of cultural research in general. This is a direct result of project management being one of the newest management disciplines, with its origins in the 1950s, and current focus on culture and its potential impacts on the project. Two keys features of this effort are particularly worth noting.

First, project management culture research is a small portion of all project management research efforts. As noted earlier, a 40-year literature survey found that 4 percent of the project management, English language based, publications were related to culture (Henrie 2005, 24). A subsequent 2005 detailed project management culture research survey confirmed that project management culture research continues but at about the same levels (24).

A notable second item is that no single or isolated group of project management cultural researchers has emerged as the leader in the area of study during the short time frame this has been a research topic. While "Many authors from many different discipline areas such as systems theory, systems thinking, management theory, sociology and project management have highlighted the criticality of [culture]" (Henrie and Sousa-Poza 2005, 5), no single person or small subset of researchers has emerged as the leader or leaders in this research field. The literature is also clear that a definitive project team cultural model has not been developed as has transpired in the national and organizational culture research areas.

In alignment with the fact that there is no single or small set of project team culture researchers, the literature also identifies a lack of a single project team culture definition. Some articles describe project team culture as the attributes of learned behavior and shared knowledge, while others fall back on national culture and organizational culture definitions, such as Hofstede's and Schein's. These approaches fail to take into account the unique and temporary nature of projects as well as the project team interactions. We will explore project management culture attributes and contributing factors in later chapters but for now we will leave the project team culture definition open.

Another common theme is, as was mentioned earlier in this chapter, culture is not something that can be directly tested or measured. These indirect measures include items like the project team's philosophy statements, its heroes, symbols, rituals, and ceremonies. Each of these is an essential and critical aspect of the project team's overall culture and provides a means of indirectly measuring or identifying the team's culture as well. The rapid formation and relatively short duration of a project team may limit the availability of these indirect measures to a small subset or a rapidly changing set of observable variables. Thus, with no direct measure of culture possible and a potential sparseness of indirect measure data and information, the researcher is challenged to identify the project team's culture based on indirect indicators.

A third challenge to project team culture research is developing an extensive data set from which to infer the entity's overall culture. Technical team culture cannot be determined by assuming that all members of the team share a common national culture or even that people from the same nation will share a single culture. In reality, every individual's culture reflects his or her unique learning experiences, social interactions, education, and the level of diversity and power distance he or she has been exposed to (Smits 2013, 21). Hofstede supports this position as he clearly states that a nation's culture index, by itself, has minimal meaning. You must have other nations' culture indexes to compare to if any real meaning is to be derived. The same is true in regard to organization cultures. One cannot study one automobile manufacturer and then apply that culture to all automobile manufacturers.

As with the nation and organization cultures, you cannot measure one project team culture and then assume each member shares this culture exactly. To an even greater extent, primarily driven by the generally temporary and unique nature of the project, the individuals will have their own social variation, team diversity, and power relation attributes. This uniqueness results in a very distinctive project team culture that may be rapidly changing.

In summary, within the area of project team culture research, at this point in time, there is no singly accepted model which the researchers can replicate. Consistent with other culture research areas there is also no single project team culture definition. Further, the challenge of measuring the project team culture is complicated by the very unique and temporary nature of project teams.

Another challenge to project team culture studies is the compounding issue of multinational project teams. Later in this chapter we will discuss some of the issues which cross-culture project team or as it is sometimes referred to, the multinational project team, investigator and team leader faces. Before that, the next section expands the difference between national, organizational, and project team cultures further.

2.6 CULTURE: NATIONAL, ORGANIZATIONAL, AND PROJECT TEAM: WHAT IS THE DIFFERENCE?

The previous section provides a short discussion on various culture theories which are based on extensive research. This discussion only touched the surface of available culture researchers and the reader is encouraged to delve into greater depths which are available for consideration. A cautionary note and a continual theme in this book is that culture is a topic which cannot be directly measured, so all analyses occur from inferred findings. As readers pursue this area of study with greater interest, they will observe conflicting thoughts, ideas, theories, and concepts between the various researchers, academics, and consulting organizations. Hampden-Turner and Trompenaars provide a sound basis for at least one of the differences that is encountered as even the study of culture cannot be culture free. It is virtually impossible to prevent the researchers' own cultural traits from penetrating their cultural research (Hampden-Turner and Trompenaars 1997, 149). So, culture research is not culture free and hence divergence and difference will creep into the efforts.

With that, let us quickly review the basis of a nation's culture. Every nation consists of different ethnic and religious groups as well as immigrants and multigenerational families. This mixture of different entities shares many different cultural artifacts such as a national flag, national song, an oath of allegiance, as well as a common language. They also share common heroes, society-accepted beliefs and values, as well as a general view of how one should interact with society, as a whole.

While the study of a nation's inferred culture variables results in the ability to define a *national* culture, it is limited in its ability to define each individual's culture. A nation's culture definition is useful as a comparison tool when looking at how other nations compare on similar scales but it is not useful when one tries to understand one individual or a group of individuals' culture or organizational cultures.

Analyzing organizational culture identifies that each entity has a unique culture and that this culture may not represent the national culture in which it is embedded. As with national culture, organizational culture is a construct which is not visible but is inferable from how the members of the organization interact within and external to the firm. Research has clearly identified that technical team culture is a legitimate concern and an area of knowledge, study, and application which is required within the modern organization and modern technical team (Morrill 2008, 15).

Organizational culture theory reflects how problems are solved as a group, decisions made, and how the employees interact as a group for the betterment of the company. As new issues occur or the need to make new decisions is encountered, the group norms, beliefs, and values guide the final process and outcomes. It is also clear that organizational culture is not static and that over time when new or different issues and problems arise or a different type of decision must be made, the organization alters or adds to its existing culture to meet the new challenges. Strong organizations have a strong but dynamic culture that evolves as the organization evolves.

Team cultures are also unique and may differ dramatically from the organization the team springs from or it may even closely resemble the broader organization's culture as well. The key to a successful team culture is the operating environment in which it is formed. Due to the project's defining traits of delivering a unique product or service, with an established budget, and within a defined, usually short, time frame the project team culture is a dynamic entity. The team must quickly form the culture foundation which establishes how it will make decisions, what the acceptable interteam member interactions are, and what its defining principles are as well as values. Until the guiding principles and basic values are established, team interaction may not be optimal and it can and often is detrimental to the overall project success.

When comparing national, organizational, and project team culture there is a distinct difference in time horizon between the three entities. National culture is one we are born into or are assimilated into. One individual or a small set of people typically does not alter a nation's culture over a short period of time. Organizations tend to have a shorter time horizon than that of the national culture. These cultures may be viewed as more fragile in that change can and does occur in shorter time frames and as influenced by specific individuals or groups. Changing an organization's culture is an objective many firms will do throughout its history. Finally, project team culture's time frame is over a very short time horizon. A team must come together and in a very short time frame develop a culture which supports the project to be successful. A single person or a small group of individuals can, and often does, have a major impact on the team's culture. This culture may be extremely dynamic with rapid changes occurring in response to internal and external events.

In this section, we have explored the relationships between national, organizational, and project team culture. The discussion identifies that each of these groups' cultures has different time horizons. It is also discussed how a single individual or a small group of people can have little to major impacts on the organization's culture. This section also identifies how culture may be very consistent to highly dynamic depending on the group.

In the next section, we step outside the single nation, single organization, and single national member project team culture environment to the setting of multinational teams and the associated cross-culture research challenges.

2.7 CROSS-CULTURAL RESEARCH CHALLENGES

Throughout this chapter, we have discussed many issues that culture researchers face in trying to determine a nation's, an organization's, and a project team's culture. The discussion of these issues has predominately been shaped around a single nation, an organization, or a project team whose members are from a common nation. Stated another way, the dialogue has been centered on a cohesive entity whose cultural roots are grounded in a common framework and environment. These groups have a common native language, heroes, philosophies, and values.

As stated throughout this chapter, there are many cohesive entity culture research challenges which researchers, organizational leaders, or team leaders must deal with as they identify and analyze the group's culture. When the organization or team includes people from other nations, each of the single culture based research and understanding challenges exists but the challenges of researching multiculture entities add another level of complexity to the process and methods of interpretation.

In this section, we highlight some of the challenges and issues which occur within cross-cultural research as it applies to teams. Within this context and as the literature identifies, "Understanding culture as it is manifested across organizations from different societies—cross-cultural organizational culture analysis—is an extraordinarily difficult undertaking, as is reflected by the relative lack of literature on the topic" (Aditya 1999, 1). As but one example, Liamputtong starts things off by identifying that "In conducting cross-cultural research, it has been found that it is rife with methodological and ethical challenges" (Liamputtong 2013, 2).

For this book, a cross-cultural or multinational team is defined as a team that includes, as predominate team members, individuals who were born and raised in different nations and as such there is diversity in primary or first languages and national cultures. While an argument can be made that teams which include personnel from different organizations, within a single nation, could be considered organizational, cross-cultural, we are limiting this book's focus to multinational project team members from different nations.

The need to understand cross-culture team interactions is becoming an ever increasing necessity. Today's global economy, high-speed and worldwide communications, rapid transportation means, worldwide company interactions, and multination collaborations create a common environment where people of different nations must work together. Successful cross-culture team interaction provides a foundation for enhancing the output and outcomes of these multinational groups while dysfunctional cross-culture team interaction hinders or even prevents the effective and efficient delivery of the intended outcomes. The literature is clear that multinational projects have failed primarily due to cross-culture clashes which were not overcome. To increase the probability of being successful, managers and teams must develop the skills required to work within the multinational group setting. As such, they as well as academic researchers are faced with the challenge of analyzing and understanding the cross-culture environment and developing models which allow the development and transformation of the multinational group into a common culture working team.

As stated earlier, cross-cultural environment based research issues all include single nation or common culture based project team problems. These issues include (a) the inability to directly measure culture, (b) generally small sample sizes, (c) the expense of conducting the research, (d) very real ethical issues, and (e) as stated, a lack of generally accepted research methods and tools. Each of these matters exists within cross-culture teams as well as other challenges such as the lack of accepted and proven research guidelines. "Although discussions of cross-cultural research have provided rich understanding of the complexities of conducting such work, few researchers have provided sufficient guidance for conducting cross-cultural evaluation research...." (Letiecq 2004, 343).

We are getting ahead of ourselves by jumping into discussions on the lack of replicable cross-culture research methods. Let us step back and look at some of the higher level issues first. At a higher level, cross-culture research suffers from a lack of a general theory to work from (Lim and Firkola 2000, 134). The lack of a general or unifying cross-cultural theory is an extreme challenge to this field of study. Without a generally accepted theory, each researcher is faced with trying to frame his or her research within a structure which allows comparative analysis between environments. Without a clear theory, it is very difficult to move forward with a universally accepted understanding of cross-cultural interactions (136).

While a general cross-culture research theory is not present, researchers continue to strive toward understanding this setting. They do so through the application of different methodological research approaches. This approach has not been very successful as the cross-culture research literature clearly identifies that, at this point in time, cross-culture methodology is fraught with issues and problems. As the literature identifies, "The majority of scholars are in agreement that methodological design in cross-cultural studies is problematic, difficult and demanding with most researchers expressing negative opinions" (Jogulu and Wood 2008, 1).

These methodological problems or dilemmas cascade down the research ladder to the actual methods which the researchers apply to this area of study. To put it into context, theories and methods are tools by which the researcher and practitioner can work from. The value, benefit, or capability obtained from these is what one can obtain from their use. Cross-cultural research lacks a unified and accepted theory as well as a divergent set of methods. The researcher and practitioner must keep these limitations and issues in mind as they proceed forward (Goethals and Whiting 1957, 441).

Further challenges of cross-culture research methods include which instrument to use, how the analysis will be conducted, what unit of analysis will be applied to the various indirect measures, and there is no universal identification or acceptance of cross-cultural specific measurement variables. While cross-cultural researchers have developed and provided various research methods, at this point in time, the research community has not adopted any one or a small subset as the preferred and accepted approach that allows comparative analysis between different multinational project teams.

In summary, in this section we have highlighted a range of cross-cultural research issues that exist today. This section also highlights that while culture research is very challenging, cross-culture research is even more challenging. Yet, while the challenges are many and much time can be spent on the various issues there is a great need to understand what and how a multinational team culture forms and the key variables on how to guide its formation and existence. By developing a better understanding of the cross-culture development process and challenges, we can provide multinational team managers with essential tools which they can use to enhance the probability of project success. This understanding also provides key information which can be leveraged to change team culture as well.

The next section briefly explores the culture change process.

2.8   CAN ONE CHANGE ONE'S CULTURE?

Whether one can change one's culture may seem like a rhetorical question at this point in the chapter as I have made numerous references to (a) the need to create a successful team culture, (b) culture as the forming of a common decision-making process, and (c) culture being a common group method of interacting. I have also made direct statements about changing or creating culture, so the apparent and obvious response to this section lead-in question is a resounding yes! Answering this question sets the stage for discussing the process of how one may change the team's culture and to ask the question: What drives one to change one's culture?

To answer these questions we put it in the general perspective that, in general, we always tend to resist change. The resistance to change is normal and based on the fact that within the context of status quo we find comfort, even if it is a less than ideal environment, as we can predict or know what is going to happen. The ability to understand and know what will probably happen in most situations brings individual comfort and a low-risk environment. To change how we do things requires us to move outside of our comfort zone and to take risks. This creates anxiety, fear, and discomfort.

Developing a new team's culture partially involves the process of reducing the team's overall anxiety level through an evolutionary path of developing a common decision-making process as well as the team's collaborative interacting process or processes. By developing common, and team-accepted, norm, philosophy, and values results in a lowering of anxiety.

This lower anxiety level is a direct result of everyone knowing what is acceptable or not, how they should interact in various situations, and how problems should be resolved. The individual does not need to think about how he or she should proceed as the way forward is already built within his or her culture. As such, the opportunity for conflict is reduced and the anxiety level is also reduced (Schein 1990, 110).

Thus, not only can people change their culture but within a new or changing team environment the anxiety of not understanding what and how acceptable decisions are made or what is or is not an accepted means of interacting within the team provides the push to change their culture, so they fit within the group or the new environment.

SUMMARY

This chapter discussed culture at the national, organizational, and team levels. We framed these discussions within the guidelines of some of the leading cultural researchers' models. Each of these profiled researchers provides a different focus on the topic with resulting different cultural models and research approaches. In some circumstances, the models overlap but each has its unique aspects.

While there are differences between the various cultural models, some consistencies exist between them. One consistent focus is that culture is learned. We obtain culture through interaction with the group and learning what is acceptable and what is not as well as how decisions are made, heroes born, and what language is used. Each of the researchers also agrees that culture is dynamic and changes over time. While the rate of cultural change and how it comes about is not universally agreed upon, they do agree that culture is subject to and does change and that the change can be purposefully produced or allowed to change based on time and environment.

This chapter also identified that if the topic is national culture, organizational culture, or team culture, there are different views and impacts according to the time component. The different magnitudes that these culture realms exist in result in a different time component and ultimately the fragility of the current culture. The chapter presents that national culture has the longest time horizon and is the least fragile of the three cultures considered. On the other end of the spectrum is the project team culture which has the shortest time horizon and has the potential to be highly dynamic.

Supporting the concept of time and fragility from within the various literature sources, there is agreement that culture must be defined at different strata. One cannot develop an understanding of a nation's culture and expect that this culture is fully replicated at the organizational level. The same applies at the organizational level as its culture cannot be defined exclusively by the nation's culture it resides within as well as the subcultures that exist within an organization. The logic continues in that a project team's culture is not an exclusive reflection of the nation and organization that the project occurs within either.

Another constant literature and research theme is that there is no direct measurement of culture. To study and identify a group's culture requires one to look at culture's indirect indicators which you can then use to infer the group's culture. These indirect measures include items like the project team's philosophy statements: its heroes, symbols, rituals, and ceremonies. Each of these is an essential and critical aspect of the project team's overall culture and provides a means of indirectly measuring or identifying the team's culture as well.

This chapter also presents the reader with several definitions of culture. While there are over 300 culture definitions, we focused on the few which the literature frequently refers to. As referenced earlier in this chapter, Dr. Hofstede defines culture as the software of the mind. Dr. Trompenaars and Dr. Hampden-Turner define culture as the way people solve problems, particularly related to relationships, time, and the external environment. Dr. Schein defines culture as the pattern of shared basic assumptions that were learned by a group. Terrence E. Deal and Allan A. Kennedy's collaborative efforts define culture around the organization's values. It is their view that values "… are the basic concepts and beliefs of an organization; as such they form the heart of the corporate culture" (Deal and Kennedy 1982, 14).

The chapter also explored the current state of project team culture research and its associated culture definitions. The outcome of this literature review is that this area of study is very new and not fully defined at this point. No single project team researcher or small group of researchers has emerged as the leaders in this field. There is also a continuation of the theme that there is no single definition of project team culture. Many culture researchers fall back on definitions developed at the national and organizational level. While this is a common approach, it does not agree with the fact that one cannot directly extrapolate one nation's culture to another or one organization's culture to another. Thus, the ability to extend these definitions to the project team may or may not be a valid approach. Further research is required to definitely identify a common project team culture definition.

This chapter also briefly looked at the challenges of culture research in general and within the context of cross-culture studies. It discussed how cross-cultural studies' challenges are inclusive of the single nation or organization culture investigations as well as having their own unique problems such as a lack of accepted theory, methodologies, and research methods.

The key takeaway from this chapter is that we exist in a cultural contextual set of relationships which are inclusive of our nation, organization, and project team. To be successful as project managers, we must understand the group's current culture, know what culture we need to be successful, and that we can affect cultural change to provide a foundation for successful project implementation. We must also understand the process which is required to implement constructive change. The following chapters delve deeper into the aspects of projects and project team cultures.

CHAPTER 3

A REVIEW OF CULTURE
IN PROJECT LITERATURE

*It must be considered that there is nothing more difficult to carry out
nor more doubtful of success nor more dangerous to handle than to
initiate a new order of things.*

—Machiavelli (2013)

3.1 INTRODUCTION

The chapter presents a discussion on project management culture in both
a homogeneous (or single nation) culture context as well as within a multinational project team environment. The objective of this chapter is to
present the reader with foundation information of culture within project
environment and to begin the discussion on what cultural differences exist
within a project team, which consists of predominately a single culture, as
well as project teams where a diversity of cultures are present.

To meet this chapter's objective, a review of the state of project management professional organizations is presented as well as a review of the
current project management culture literature. The project management
professional organization review expands the discussion in Chapter 1 by
reviewing other project management professional organizations' history,
stated charter, and their views on project team culture. In this process,
this chapter provides a comparison and contrast between the various
project professional organizations' views on project management and,
specifically, project team culture.

Following the discussion on project management professional
organizations, the chapter delves deeper into the current state of the

published project culture literature. This review is inclusive of homogeneous national culture research as well as multinational project team research. A homogeneous national culture encompasses those individuals from within a single nation who have a common culture. Multinational project teams, on the other hand, exist when individuals from different cultures, sometimes referred to as cross-cultural, are combined into a project team. This discussion extends the information discussed in Chapter 2.

The intent of this chapter is to help the reader formulate ideas and concepts about analyzing a project team's culture by means of the combination of references provided and the information contained in this and Chapters 1 and 2. This understanding then provides the practitioner foundational knowledge on how cultural change may be planned and implemented such that the resulting team culture achieves maximum effectiveness. The summary of this effort is that the student of team culture can fully understand the points that (a) all teams have a culture, (b) the project team culture may support or hinder successful implementation of the project, and (c) unless the project team has an effective culture it is extremely difficult for them to achieve project excellence.

The next section presents a review of the history of several project management professional organizations and publicly provided literature which refers to or incorporates culture.

3.2   THE PROJECT MANAGEMENT PROFESSIONAL ORGANIZATION

Project management professional organizations can be classified into a few basic forms. First, there is the stand-alone project management professional organization which has a central office and all activities, certifications, and policies originate from this central group. The Project Management Institute (PMI) can be classified under this area. As found on the PMI website a 15-member volunteer Board of Directors and the Executive Management Group exist and are responsible for guiding day-to-day operations. From this central governance group local chapters are formed based on geographic location and the desire of the local members to establish a local chapter. The local chapters provide a means for individuals, who are located in the same geographic area, to meet, exchange ideas, enhance their skills, and share experiences. These chapters provide social, educational, and job reference opportunities. The local chapters are entities which have been approved by the PMI central organization—that is, they are not independent of PMI but abide by all PMI local chapter requirements.

The second form of a project management professional organization is a federation-based organization. A federation organization is a union of partially self-governing organizations under a central governing body. As such, and in alignment with the preceding sentence, a project management professional's federation organization consists of member organizations which are loosely combined under a central entity. The member organizations have certain rights and abilities to act independent of the central organization but, by agreement, the central organization reserves and holds specific rights and abilities to itself. "In a federation, the self-governing status of the component states [in this case the member associations], as well as the division of power between them and the central government [in this case the central organization], are typically constitutionally entrenched and may not be altered by a unilateral decision of either party …" (*Wikipedia* 2013).

An example of a federation-based project management association is the International Project Management Association (IPMA). The IPMA parent organization is the overarching structure which represents a global set of member associations (IPMA 2005).

As the governing body of this federation, IPMA provides many functions such as being the global representative of all member associations, developing and publishing project management standards, guidelines, and best practices, and it provides leadership and support in promoting and developing the project management discipline and professionalism (IPMA 2005). IPMA also manages the project management certification process but each member association establishes the organization which performs the actual assessments and certifications. In this process the member associations have the liberty to adapt "…some factors and requirements to their local needs" (IPMA 2005). The ability of the member organizations to modify or adapt their individual certification process is an example of how power is shared between the central organization and its member organizations.

The third type of project management professional organizations is the member associations which are part of the federation. Member associations are nation-specific organizations which focus on their geographic locations to develop project management competencies within their unique cultures and in alignment with IPMA's overall guidance, decision-making processes, and governance structure. These member organizations focus on the needs of the members within their national borders, in alignment with the IPMA's general guidance and reserved roles. They have a high level of autonomy and self-governance but not unilateral autonomy and self-governance that a fully independent organization would have yet, the level of autonomy and self-governance exceeds that provided to local chapters under the central governed professional organization. The Association for Project Management (APM), the American Society for the Advancement of Project Management (asapm), and the Australian Institute for Project Management are just three examples of IPMA member association organizations.

A fourth type of professional organization is identified which provides information and services for members who are interested in project management. These organizations tend to be industry specific and have a tendency to focus on a subset of the overall project management profession. As but one example is the Association for Project Managers organization. The Association for Project Managers mission statement is "To promote project management excellence in the design and construction industry through knowledge sharing, education and quality management" (APM 2013). These organizations generally do not have local chapters or member associations. They also tend to focus on a subset of the overall discipline rather than the full discipline spectrum.

In summary, there are several different project management professional organizations across the globe. Each type has a unique structure and approach to advancing the project management profession. Part of the society's efforts include (a) establishment of codes of conduct and codes of ethics that all members in good standing will follow, (b) providing a means for members to interact on a common subject, (c) the support of educational opportunities, (d) establishment and development of industry-specific standards, and (e) being the discipline's advocate at the national and global level.

Depending on the organization type, the level of explicit inclusion or exclusion of culture knowledge and information varies as well. The more global organizations tend to include culture within their standards, recommended practices, and articles while the industry-specific organizations may not.

3.3   PROJECT MANAGEMENT PUBLICATIONS

As a key focus of this chapter is on the state of project team culture, as discussed in the various forms of published literature, this section highlights some sources where this information is derived. When performing a project team culture literature review, many different avenues (the types of publications) are available to practitioners, academics, and researchers. These various publications range from books, newsletters, and monthly magazines to peer-reviewed (and often academic) journals. As a note to the reader, this book does not include newer technology based information sharing sites such as blogs, wikis, and other social media sources. We leave these information sources to the reader to pursue and investigate.

In alignment with this book's exclusion of social media information sources as sites which have a specific focus and audience, there is also a distinct difference between the classification of popular or general interest publications, such as newsletters and magazines, and scholarly journals which are also referenced as peer-reviewed or academic publications. The general interest or popular magazines are characterized as literature sources which are intended to inform or entertain the audience. These publications rarely cite other works; predominately they are not based on detailed research efforts, they also generally fail to include supporting data such as statistical analysis. The articles are generally not peer-reviewed, either. That is, the articles may receive an editorial review prior to publication but the information provided has not been reviewed by leading experts in the field as a vetting process. The focus of these articles is to provide practitioners and interested individuals insights into current happenings, and general information, versus educate or as commonly stated in the academic realm, "contribute to the body of knowledge."

Scholarly journal articles are the converse of general interest articles, the focus of which is to keep practitioners informed of what research is being performed in their field of expertise and to "contribute to the body of knowledge." The articles are based on either basic research, applied research, or a combination of research processes. Basic research is also called pure research or fundamental research. The objective of basic research is to develop a fundamental understanding of the topic. This is different than applied research. Applied research involves the scientific study of "real-world" events which are designed to develop a better understanding of practical solutions within the discipline. Applied research wants to answer real-world problem questions while basic research wants to advance the body of knowledge, not specifically to advance or propose answers to real-world problems.

Published basic and applied research articles must have supporting references and the frequent use of statistical analysis is encountered. Prior to publication these articles are peer-reviewed by experts in the field. These articles provide the practitioners insight into concepts, ideas, tools, techniques, and methods which have worked and have not worked.

With the distinction between popular and scholarly journal articles explained, the following sections provide a highlight of where the interested reader can find information about project management and project team culture.

3.4 LITERATURE SOURCES

In the area of popular or general interest publications, the project management discipline literature sources provide access to several newsletters and magazines. Over time these sources evolve, merge, and disappear but at the time of this book the following popular interest literature sources were all very active and available in various formats.

- *Project Manager Today*—This is a monthly magazine that members of the Project Management Specialist Group (PROMS-G) have access to. This magazine's aim and scope is to address the project and program management community within the United Kingdom (UK) and countries external to the UK.
- *PM Network*®—PMI publishes this magazine monthly. Its aim and scope is to keep the project management professional "…updated on the latest tools, techniques and best practices…with real-world information you can apply to your work and career" (PM Network 2013).
- *PMI Today*®—The PMI Inc. also publishes this monthly newsletter. The newsletter is intended to provide its members with up-to-date information on institute news and events as well as volunteer opportunities ("Our Publications" 2013).

For the academic and the researcher, who strive to publish in peer-reviewed journals, the project management profession provides several avenues to achieve this objective. The various project management peer-reviewed journals include:

- *International Journal of Project Management* (IJPM)—"The APM and IPMA collaboratively publishes the IJPM eight times a year. IJPM articles cover a broad range of in depth articles across the full project management basic and applied research spectrum" (IJPM 2013).
- *Project Management Journal*® (PMJ®)—The PMI Inc. publishes this peer-reviewed journal on a quarterly basis.

The peer-refereed academic and research publication of PMI, …
features state-of-the-art management techniques, research, theories and applications. It addresses the broad interests of the project management profession. (PMIJ 2014)

While the IJPM and PMJ literature sources are dedicated to publishing
project management specific articles there are many other scholarly journal publication sources which often include project management research.
Many of these other sources are associated with a specific discipline. As
an example, The Institute of Clinical Research magazine, CRFocus, has
project management articles which focus on managing projects within the
area of clinical research.

Another academic-level journal which frequently publishes project
management research articles is the Engineering Management Journal
(EMJ). EMJ is published four times a year by the American Society for
Engineering Management. As stated in EMJ's editorial mission "EMJ is
designed to provide practical, pertinent knowledge on the management of
technology, technical professionals, and technical organizations" (ASEM
2013). Then there is the International Electrical and Electronics Engineering (IEEE) association. It is the world's largest professional association
that is associated with the advancement of technology. Within the IEEE's
professional association efforts is the development of standards and publication of research articles. These efforts are inclusive of engineering
management and project management activities, processes, standards, and
research.

With this introduction to various project management literature
sources, the following section discusses the current state of project team
culture as presented in these and other sources.

3.5 PROJECT TEAM CULTURE LITERATURE REVIEW

This section provides a review of project team culture as presented in
project management books, general management literature sources, popular magazines, project management standards, as well as peer-reviewed
journals. The objective of this section is to provide the reader a deeper
understanding of project team culture and an understanding of how the
discipline's view of culture has evolved.

This section begins with a look at how culture research within projects has evolved over time as it is referenced within project management standards. To start this discussion we look at the evolution of culture which has occurred through the five editions of the Project Management Body of Knowledge (PMBOK®).

In August 1994, an Exposure Draft of *A Guide to the Body of Knowledge* (PMBOK) was released for review and comment. This Exposure Draft covered the topic of culture within the context that each organization has a unique culture in and of itself. These unique cultures reflect the organization's specific set of norms, shared values, beliefs, decision-making processes, and expectations. These cultures also reflect and support the organization's specific views on authority. The superset of organizational culture impinges on and influences the project team cultures within its structure (PMI 1994, 9).

In 1996, the second PMBOK edition was published. In the second edition's Section 2.3.2, project culture continues to be discussed within the context of a unique organizational culture. There were minimal changes between the first PMBOK edition and the 1996 PMBOK edition. The view that the organization's culture directly influences and shapes the project team culture remained (PMI 1996, 18).

When comparing the first and second editions, one major change is noted. Specifically, the general cultural view changes from being that of a potential constraint to a potential direct influence source. This implies a change in thought that culture can have either or both a positive and negative influence within the project team.

The third PMBOK edition was released with an expansion of cultural consideration into several new sections. To highlight these new inclusions, the review starts with a new addition in Section 1.3. This new inclusion identifies that culture is included in the management of project by project schema of Management by Project. Management by Project is an organizational process which applies project management tools, techniques, and methods within a routine maintenance, fabrications, or assembly-type organization. Within this schema the organizational culture is closer to the project team culture than the converse view in the first two editions (PMI 2004, 8).

Culture was also added to Section 1.5.3 where it is stressed that the project manager needs to be aware of and specifically examine the organizational culture (PMI 2004, 14). The third edition also adds culture as a discussion topic in regard to organizational project management system maturity as well as expands Section 2.3.2 to include specific factors in which culture is reflected. Finally, culture was added as an input source in Scope Planning, Human Resource Planning, and Recognition and Rewards sections.

The PMBOK fourth edition continued with the expansion of culture even further. In this edition, culture shows up almost immediately in Section 1.11 in the acknowledgment that project team members come from a diverse set of cultures and backgrounds (PMI 2008, 4). Culture shows up again, in the fourth edition, in Section 1.8 with the discussion on Enterprise Environmental Factors (14). The previous edition's discussion about culture in Management by Project organizations no longer exists.

The fourth edition also includes culture references in:

- Section 2.2 Projects vs. Operational Work
- A continual reference to culture in the Organizational Cultures and Styles section
- Develop Project Management Plans Inputs
- Direct and Manage Project Execution Inputs
- Develop Human Resource Plan Inputs
- Chapter 9—Project Human Resource Management
- Chapter 10—Project Communications Management
- Chapter 11—Project Risk Management, as well as
- Appendix G

The PMBOK fifth edition was released in 2013. The fifth edition greatly amplifies the utilization of culture throughout the book. Culture is specifically referenced at least once in nine out of the 10 knowledge areas as a key environmental factor. This is the highest knowledge area inclusion of all editions. The fifth edition is also the first time that a cross-cultural concept is specifically identified as well.

In summary, as shown in Exhibit 3.1, the PMBOK has steadily increased the use of culture in a descriptive form as well as key environmental inputs. While culture has been a topic since the first edition, the total frequency of occurrence has increased about 7.6 times, from seven occurrences in the first edition of the PMBOK to 53 references in the fifth edition of the PMBOK. Also, the number of sections where culture appears increased by a factor of 15. That is, the first edition included culture in two sections while the fifth edition expanded this to 30 sections. A notable and significant expansion also occurred in how culture is included in the specific knowledge areas. In the first edition, culture was not included as a key environmental input. The second edition had culture as an input in one knowledge area, the third edition increased the references to three out of the nine knowledge areas, the fourth edition increased the knowledge area input references to four, and the fifth edition included culture as a key input consideration in nine out of the newly expanded 10 knowledge areas.

The significant increase in culture reference is linked to the various disciplines, such as engineering management and project management, recognizing culture impacts. There is a linkage between the discipline's recognition of the importance of culture and the occurrence within the various publications. Exhibit 3.1 reflects the increased culture or cultural reference frequency according to total count, total number of document sections where culture or cultural activity occurs, and the number of knowledge areas where culture specifically occurs.

The identified rate of change graphically takes the shape of a slightly positive exponential curve. This can be viewed as an indicator of how culture is becoming a key consideration that influences projects. The fifth edition's inclusion of cross-culture discussions also fits within the greater body of culture research which identifies that cross-culture, multiculture, or multinational projects are distinct from a more homogeneous nation cultural environment

It must be noted that in all PMBOK editions culture is predominately discussed in the context of organizational culture and the effects and impacts that an organizational culture may have on the project team. What is noticeably absent is the discussion or reference to project team culture as a unique organization subculture.

The one notable exception to the PMBOK's organizational culture predominate view occurs in the fifth edition, within the section titled *Develop Project Team*. This section presents team culture as a distinct discussion point for the first time. As a specific reference there is acknowledgment

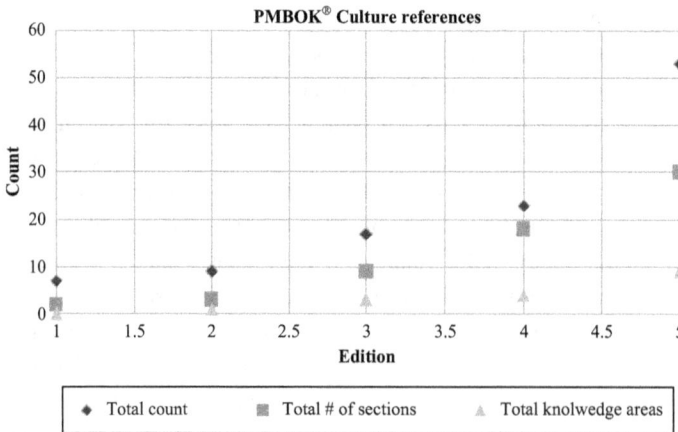

Exhibit 3.1. Culture literature reference frequency.

that a project team culture is unique. The section also goes on to identify the importance of creating a positive team culture which supports a highly productive team that is characterized by positive team spirit and cooperation (PMI 2013a, 274).

While the PMBOK does not delve into the influences and potential affects that are associated with national, organizational, and team cultures, this is not consistent with other literature sources.

In analyzing various literature sources beyond the PMBOK, there is a clear and consistent interweaving and mixture of national culture, organizational culture, multinational team, and project team culture literature discussions. These discussions are prevalent and consistent within the literature.

Other key attributes for project culture literature include the following aspects:

1. Project teams can and do have their own culture or, as it is sometimes referred to, a subculture.
2. There is no *standard* project team culture. Each project team culture is distinctive as it is based on a different collection of project team skills, areas of expertise as well as the specific project context, level of internal and external competition, problem resolution needs, decision requirements, values, attitudes, and internal and external influences applied to the team.
3. "...cultural values influence knowledge sharing behaviours by shaping patterns and qualities of interaction needed to leverage knowledge among individuals..." (Wiewiora et al. 2013, 1164).
4. Projects are more successful if they have a culture of trust and that culture can be a unique and continuous competitive advantage source (Zheng, Yang, and McLean 2010, 765).
5. "Successful projects create their own unique culture..." (Aronson, Shenhar, and Patanakul 2013, 37).
6. Project team functional change can be difficult to achieve due to cultural influences.
7. A team's culture will affect its effectiveness in communicating and interacting internally to the project team and to external entities.
8. "...cultural changes must be supported by both primary culture-embedding mechanisms and secondary articulated and reinforcement mechanisms..." (Eskerod and Skriver 2007, 116).
9. Projects which include personnel from different cultures find that the complexity of culture interactions can be a significant barrier to communication.

10. Multinational project teams' cross-cultural challenges, issues, and potential benefits are a complex environment.
11. Culture awareness and knowledge is essential for firms to be successful in multinational projects.

This list highlights some of the common project team cultural themes. While the list is not all inclusive it does provide a sound basis of what academics, researchers, and practitioners are studying (because they see importance or value in understanding the issue) and writing about (which indicates the importance or value assigned to the topic by readers and experts in the field).

To expand slightly on these points, the following discussions consider project team culture in and of itself.

Within this context, a project literature predominate theme is that the project team culture is the organization's culture. The vast majority of literature consistently refers to, compares, and contrasts project team culture in the light of organizational culture or even national culture. Yet, there is a growing body of literature which is moving away from this focus as it identifies and discusses the concepts of organizational subcultures.

Subcultures are groups, such as the project team, within the organization that are sufficiently different that they have their own culture. These subcultures are often recognized by artifacts which are distinctly different than the organization or even other project teams within the organization. As example, one project team will have an observable artifact that the team concludes the day over a glass of wine. This informal meeting provides a relaxed atmosphere to review the day's progress and review the next day's activities. A different project team, within the same organization, culture does not support a similar informal and relaxed meeting exchange but exists within a culture of command and control versus collaboration. Rather than have an informal meeting this team artifact is one of very formal meetings, with agendas, action items, and observation of hierarchical reporting relationships. Each project team exists within the same organization but has developed its own subculture of norms, values, communication methods, and decision-making processes.

What this means is that organization subcultures can and do occur within the organization when distinct groups of individuals work in a collaborative environment. During this shared experience the group develops a shared problem-solving process, established accepted set of norms, common values, attitudes, and a socially accepted decision-making process. These socially accepted processes, norms, language, values, and basic assumptions are sufficiently unique, to the organizational subgroup, so that they separate, differentiate, or distinguish its culture from the more general organization culture. An example of this is when you overhear someone explain how it is almost impossible to understand those engineers as they speak a different language and they have their own way of doing things as well as making decisions.

What this growing body of subculture literature identifies is that teams have their own set of conflicts, decisions, communication issues, and social interactions from which the team must develop socially accepted cultural attributes for it to be successful. As Kendra and Taplin's research derived "Analysis of the stories shared by the participants confirms that a project management culture exists" (2004, 40). In this case, the project team's culture was not *fully* in alignment with the larger organization.

It is also a key research finding that, in general, different project teams do not share a common culture. That is, when evaluating or investigating different project teams each will exhibit a distinctive culture. No two project teams will have exactly the same culture. This view is referred to as an emic current versus etic current perspective.

What emic and etic perspectives refer to is how one approaches the understanding of an issue, in this case the project team culture. One concise definition of emic and etic perspectives is that:

> ... there are two long-standing approaches to understanding the role of culture: (1) the inside perspective of ethnographers, who strive to describe a particular culture in its own terms, and (2) the outside perspective of comparativist researchers, who attempt to describe differences across cultures in terms of a general, external standards ... emic and etic perspectives, respectively. (Morris et al. 1999, 781)

The emic view postulates that there is no way to make a direct comparison between two cultures due to the very differences of the cultures (de Bony 2009, 781). As is discussed later in this chapter, there are many reasons why project teams' cultures are emic in nature.

An example of emic research is identified in an article by Lung-Tan Lu. In this article, emic is clearly defined and applied as a research tool which "... attempts to identify culture-specific aspects of concepts and behavior, which cannot be comparable across all cultures. Emic researchers assume that the best way to understand a culture is as an integrated whole" (Lu 2012, 109).

An etic example would be the work of Dr. Hofstede. His research provides a comparative analysis of each nation based on an established culture indirect variable set which is intended to characterize a nation's culture. The result is a set of index values which allow one to compare one nation's culture to another's. Dr. Hofstede is very clear that his index results can only be applied when used to compare between nations, not how the nation's individual's specific culture is or a comparison of various individuals' cultures within that nation. Many of the project management culture researchers follow this analysis perspective as well.

As noted earlier there are many factors which contribute to differences between project team cultures. These differences, in and of themselves, do not indicate one culture is the absolute best way to implement a project team culture or that one is necessarily better than the other. The difference in cultures identifies that each project has a distinct context and a unique environment which results in a project team specific culture.

One factor, which contributes to project team or engineering team culture differences, is the various sets of management tools, techniques, and associated activities to which they are applied. "As the management activity is made by people who are very much influenced by their values and beliefs, no management activity can be 'culture-free'" (Bredillet, Yatim, and Ruiz 2009, 183). This clearly indicates (and is supported by earlier statements within this book and which is elaborated on further in later chapters) that the project manager or the engineering manager is an essential element in the formation and sustainability of the team's culture.

As an example, managers who have a high power distance index, that is, one whose culture is that of a stringent social class structure, will take actions, make decisions, and directly support as well as foster a project team culture which conforms to this team structure. Conversely, a manager with a low power index culture background which lacks a strong social class value, norm, and attitude will tend to foster a team environment which is more open to bidirectional communications as well as foster a lower authoritarian command and control style.

Another contributing factor to each team's specific culture is the physical, social, and technical environment in which the project is implemented. This multifaceted environment shapes the various decisions which must be made, and provides a unique set of issues which the team must resolve. It is also fully possible that the project team members have never worked together within a somewhat similar context. As such, the project team must establish norms, values, and basic assumptions which are acceptable to the project team as they make decisions, resolve issues, and communicate both internally and externally to the project team.

The merging of individuals into a culturally homogenous project team is challenged by many things. Some such areas, which impact the development of a common project team culture, include knowledge transfer, team communications, and team success. It is these types of processes which make changing or implementing a team culture change difficult within a homogeneous environment and even more difficult within a multinational team culture. The increased difficulty of multinational team culture development is directly associated with the diversity of cultures within the team. While many of the homogeneous and multinational team cultural issues are similar, the multinational interaction increases the cultural transformation significantly. The following paragraphs provide further information on each of these contributing factors.

As noted in the previous list, project management culture research also indicates a clear linkage between the project team's knowledge transfer capabilities and its culture. The foundation for this exchange is rooted in the team's underlying basic assumptions. As identified in research, "Basic assumptions, like theories-in-use, tend to be nonconfrontable and nondebatable, and hence are extremely difficult to change" (Schein 2004, 31). Thus, if the project team's basic assumptions are in conflict with the need or intent of fostering effective knowledge transfer this becomes a barrier to success. A frequently encountered example of this is the "not developed here" so it is not a good idea, process, and so forth basic assumption and the acceptance of an external knowledge set. What this means is that the project team has a basic assumption that their environment is so unique that external input, which is knowledge, is not applicable. Therefore the basic assumption rejects what may be some important new knowledge. To implement a successful knowledge transfer process requires all those involved, including the project manager and senior management, to work cohesively to ensure that both the organization's and team's basic assumptions are taken into consideration in the knowledge exchange as well as the approach to how the knowledge will be transferred (Eskerod and Skriver 2007, 118).

Culture factors which contribute to the impediment of an effective knowledge transfer include:

a. The temporary nature of the project team association. As culture is a learned process which involves the team members' direct interaction in solving issues, making decisions, and communicating, there is a time component in this process. With short-duration projects the time component hinders development of a cohesive culture. This can directly hinder knowledge transfer within the project.

b. A rapidly changing team and knowledge environment. Projects are involved in development and implementation of new and unique endeavors. While the project team members may have a core or in-depth understanding of the technology and processes, each project environment is distinctive and new knowledge is being developed. This can result in challenges to the team's basic assumptions as they were previously developed in a different context.

A continuation of the project management culture literature theme is that internal and external communications are essential for project success. Delivering the final output is insufficient if communication has not occurred. As the literature notes, "... project team communication is affected by the individual member's culture.... [and] the greater the diversity of individual cultures, the greater the potential for unsatisfactory communications" (Henrie 2005, "Introduction"). This cultural issue can cascade into difficulties across the project which may negatively influence the project outcome.

There are other reasons why culture inhibits or enhances project team communications such as the "Number and variety of interfaces between project and other organizational entities. In the same way that a large number of different disciplines on a project can create a management challenge, a large number of different organisations can as well ..." (GAPPS 2006, 5).

The literature also identifies how culture is the very foundation of communications. As different cultures interact and if they are not aware of their cultural disparities, this can then become a major source of *noise* which can interrupt or distort the message's intended meaning. Problems arise when communication receivers attribute meaning to a message according to their own cultural frame of reference rather than that of the sender. The different reference points may easily result in miscommunications or a reduction in effective and efficient team communications (Loosemore and Lee 2002, 518). Failing to effectively and efficiently communicate greatly reduces the project team's ability to be successful.

The project management literature also contains a significant amount of information on the relationship between the project team attributes of (a) a high performing project team, (b) a fully functioning project team, and (c) project team success. What is clear is that the project management literature identifies how important organizational and team culture is as well as the direct interrelationship between these cultures and project success.

As an example, the literature identifies that an open culture project team interaction supports project success. These team open cultures foster the capability of openness and information sharing as well as transparency

(Killen and Kjaer 2012, 557). The converse is also valid that a closed culture project team fails to effectively and efficiently support a cohesive and highly functional team.

The concept that culture is important to a highly functional and successful project team viewpoint is consistent with general management culture literature. Within both bodies of knowledge research and practitioner experience the data identifies that culture is a key component. The literature also clearly links organizational management and project management failures to the lack of a common culture or divergent and competing cultures.

As there is some commonality between general management and project management literature, a natural question becomes what and how is general management culture and project team culture literature related? The next few paragraphs expand on the following answer but the overarching perspective is that general managers and project managers utilize many of the same skills, tools, and techniques in performing their respective roles. They also face many of the same cultural issues. In this regard, the general management literature provides a source of information which adds to the overall project management culture literature.

To expand on the previous paragraph, from an overall perspective, the general management body of culture-based literature is much larger than the corresponding body of project management literature. This body of knowledge includes extensive research on cultural impacts within the management discipline. Additionally, general management cultural research has a longer history than project management, as it dates from the early part of the nineteenth century, which is around the 1930s. In this time frame the researcher, Elton Mayo, who was assisted by W. Lloyd Warner began to study organizational workgroup cultures (Park, Ribiere, and Schulte 2004, 106). These initial general management research efforts predate project management research since project management, as a discipline, originated in the 1950s.

A key takeaway from the general management body of literature is that culture is either a significant contributor or a significant barrier to management success. One approach to understanding cultural management impacts is by looking at the organization from a viable system perspective.

The viable system organizational capability literature describes companies as existing within seven environments. These environments consist of (1) a commercial environment, (2) technical environment, (3) economic environment, (4) political environment, (5) social environment, (6) educational environment, and (7) an ecological environment. Of these seven environments, culture plays a key part in all with the exception of the ecological environment. This is due to the fact that the ecological environment is the foundation and supporting infrastructure for the other six environments (Christopher 2007, 360). Clearly culture can and does have a significant influence on general management ability to be successful in the management of the company's *soft* structures.

In respect to a company's soft structure and what that refers to, the general management literature identifies how companies consist of hard and soft structures. The hard structure includes items such as the physical location, formal structures, information systems, process systems, and overall corporate strategy. These items tend to be very difficult to change once put in place. That is, they are hard points within the overall organization's structure. Soft systems, on the other hand, include human resources, leadership skills, management skills, communications, the firm's values, norms, and attitudes. These items are not as physically tangible as hard system components. Additionally, when reviewing those organizational parts, which are deemed soft systems, it is easy to see that culture is associated with each of them. This is directly related to the fact the organizational human element is directly associated with each item and humans come with a culture.

These general management soft systems are directly comparable to the project manager's roles, project structure, as well as how the project is implemented. That is, the general management soft systems are also found in the project management team structure. This is further support that many of the aspects, findings, and applications of general management culture research can be applied to the project management culture body of knowledge.

While project management research continues to expand this discipline's body of knowledge academics, researchers and practitioners can leverage general management literature to make well-grounded assumptions as to the potential impacts culture may have on projects and the project team. As noted in the previous paragraphs, there is direct comparability between many aspects of general management and project management as they both require the knowledge and utilization of general management. They also use similar tools, techniques, and processes.

The project management culture literature also identifies another constant theme between the various levels of culture, which is at the national, organizational, and team levels. Specifically, the literature discusses how difficult it is to change a firmly established culture. This finding is in full congruence with the other bodies of culture literature which stresses how difficult it is to successfully make such a change.

Another area where there is agreement between the various management disciplines' literature is that while implementing a culture change is difficult, conversely, implementing a broader project or organizational change is either resisted or assisted by the group's culture. The literature identifies that change is difficult at any time. Yet, the level of difficulty only increases if the intended change is not in alignment with the organization, project team, or other groups' basic values. There is a higher probability that organizational change will be successful and occur at a faster rate if cultural change is included and the core values are maintained (Karlsen 2011, 244). The research is clear that culture has a broad and lasting influence on organizations and project teams.

3.6 MULTINATIONAL PROJECT TEAM CULTURE LITERATURE REVIEW

In this section, the homogenous national culture literature discussion is extended to include multinational or cross-national culture literature research. Depending on the literature source the reader will find reference to both multinational and cross-national cultural research. Multinational research tends to be defined as research which involves entities from two or more countries or nations. Often multinational is used as a descriptive term as in a multinational company which means that the company has presence in more than one country. Cross-national research is similarly defined as research that pertains to or involves at least two nations. Cross-national research can be viewed more as an action phrase as in performing research which involves personnel from two or more nations within a single context. While a slight difference can been seen in these views it is not uncommon to see these terms used interchangeably. For this document both phrases are used while discussing organizations and teams which include people from more than one country or nation.

Cultural research, within entities that include more than one nation, as presented in *Multinational Project Team Communications: International Cultural Influences*, continues with the theme that cultural research roots arise in anthropological studies. Further, the research identifies that each member of the organization or team culture may have a unique set of underlying beliefs, values, norms, and basic assumptions (Henrie 2010, 2).

Cross-national cultural research, by description, is an expansion of homogeneous national, organizational, and project team culture research. This expansion focuses on the interactions of different cultures within the various settings versus the study of culture within a homogeneous background. Support for the differentiation in homogeneous and cross-national research arises from the management research area. Specifically, management research, rather than project management research, identifies that cross-national cultural management studies involve organizations and people who live and work in other nations and in other cultures (possibly within the same nation). Cross-national research is efforts which involve situational contexts which include people from different nations and cultures working together within an organization or team environment (Adler 1983, 226).

Before delving too deeply into the cross-national cultural literature discussion it must be identified that researchers utilize a range of different terms, such as but not limited to multinational and cross-national, to describe the cultural interaction which occurs when people of different national cultures come together to perform some common task, activity, or project. Various researchers refer to this environment within the context of (a) multicultural, (b) global, (c) intercultural, (d) international, (e) transnational, (f) intercultural, (g) cross-cultural, (h) cross-national, and (i) multinational. In general, regardless of what the researchers or authors label this environment, they are referring to an organization or team, within the context of this book, which consists of people who are working together but do not share the same national or ethnic culture. The divergence of cultures sets in motion the expanding complexities of culture within this multinational culture discussion.

A review of cross-cultural project management literature identifies commonality between the types of issues which occur in homogeneous and multinational cultures as well as the potential benefits which can occur between these different project team types. The literature also identifies that somewhere between 25 and 50 percent of the multinational team attitudes can be explained by the team members' national culture (Gannon 1994). As discussed in the following paragraphs, while there is a commonality between homogeneous national and multinational cultural issues and potential benefits each of these tend to be expanded within multinational project team environments. The expansion of team issues and the scope and magnitude of potential benefits can be linked to complexity of the environment. The complexity of project teams can be partially explained by complexity theory, which is discussed later in this chapter.

We start the multinational team culture discussion by looking at its distance factor or by another description: virtual versus colocated teams. By way of background and within the context of this book virtual project teams are those teams where the team members are physically separated by various distances and have minimal to no direct physical contact. What this means is that the project team members almost always or exclusively interact through electronic means such as e-mails, teleconferences, phone calls, and Internet-based meetings. The team members are real, the project is real, yet the team rarely if ever comes together in one place at one time. Conversely, colocated teams exist when the team members share the same physical area and predominately interact in face-to-face processes such as meetings, ad hoc encounters, and *just* dropping in to see how things are going.

Understanding the team context is a key contributor to multinational project team culture discussion. The team's physical, social, and working environments and their impacts and influences the team members interaction efficiency and effectiveness. This variable's primary question involves identifying if the project team is colocated in a single physical area or the team is predominately dispersed as in a virtual team. Hypothetically, if the multinational team works within the physical constraints of a common area they have a greater opportunity to develop a mutually satisfactory culture. This occurs as the team members directly interact frequently and have similar experiences at the same time.

Converse to this concept is the idea that virtual multinational teams will find it more difficult to establish a common culture, for short-term projects, due to a lack of direct interaction and mutual sharing of events within the same context. This hypothesis is based on the fact that culture is a group-shared learning process which occurs through *direct* interaction, problem solving, and decision-making efforts. The greater the physical distance, which is the distance factor, between the team members, the higher the level of difficulty in sharing experiences, direct collaboration in problem solving, and norms. The lack of direct involvement results in less opportunity to develop a shared culture.

Cross-culture project team literature provides further information on the distance factor which supports the hypothesis stated in the preceding paragraph. Specifically, the distance factor strives to answer the question: "Is a virtual project team more culturally challenged than a colocated multinational project team?" The cross-culture literature postulates that, "Distance [that is, the team is not colocated] and culture are perceived as two aspects critical to team effectiveness in a global context ..." (Connaughton and Shuffler 2007, 389). One literature theme, along these lines, is that with physically divergent or virtual team environments there is a lack of organizational or group context and experience sharing. This lack of common context and experience sharing hinders the development of an effective culture (Hartman and Guss 1996) which cascades into the potential of negatively impacting the project team's effective and efficient project delivery.

Another cross-cultural influencing factor is the complexity of the multinational project team. At a fundamental level (and an idea which appears in the complexity literature) is that complexity is associated with anything that is difficult to understand. Complex systems are associated with nonlinear, dynamic organizations which have a feedback mechanism. Complexity is different than a random event or random system in that a complex system has regularities but these regularities cannot be fully defined or results predicted (Axelrod and Cohen 2000). Random systems, on the other hand, exhibit no form of regularities.

While there is no definitive definition or set of final characteristics which clearly define what a complex project system is, there are many variables which provide insight into what influences the system's complexity. One variable is the number of interconnecting parts. In complexity science there are references to the fact that as the number of interconnections increases the potential complexity of the system increases as well. As an example, if two people are interacting then there is a single communication channel. If a third person joins the conversation then the number of communication channels increases to three. Increasing the number of participants to four increases the number of communication channels to six. Exhibit 3.2 demonstrates that as the number of participants, as in the project team, increases the result is a positive exponential curve. This curve is mathematically defined as $(n*(n-1))/2$.

If you are part of a relatively moderate sized project team, say of 20 people, there are 190 possible communication paths. This results in a communication network with a tremendous number of communication paths, feedback loops, external energy sources, and interconnections which can alter the project trajectory.

The literature also identifies that complexity is associated with the introduction or inclusion of the human element as part of the open system.

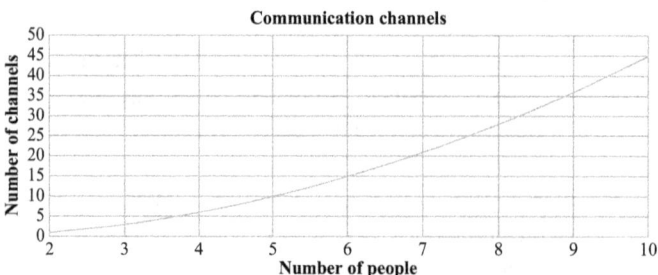

Exhibit 3.2. Communication channel example.

System science defines an open system as a system which continuously interacts with the external environment. An open system receives input or energy from external sources and provides output or energy to the environment. As stated, an open system does exchange energy from its external environment. A car engine is an example of an open system. It receives energy in the form of oxygen and fuel which is converted into work. The work produced is used in the external environment. As long as the external inputs continue the engine will continue to produce work; of course, this is under the assumption that no mechanical failure occurs. This is the opposite of a closed system which does not interact with the external environment by either receiving input or providing an output. An example of a closed system is a pendulum contained within vacuum with no means of receiving any external input such as someone pushing on it. In this example, the pendulum never changes position or outputs any work. The encased pendulum is a closed system as it neither receives nor exchanges energy external to itself.

When considering the project team and open systems, the human element is paramount in both. That is, the human element forms the project team. The project team becomes an open system as the human element continuously interacts with the external environment by receiving energy in and producing work out. This open system concept is valid for both homogeneous as well as multinational project teams.

This open system interaction and communication network underlies the complexity of the project team and interacts with the cultural development, sustainment, and change. When the project team expands beyond a homogeneous national culture, to one which involves multiple national culture team members the complexity of the system increases as well. This is demonstrated in the area of multinational project team communications. All communication is a process which involves an originator who encodes information within a message stream. The encoded message is passed through a communication medium such as the air, e-mail, or other electronic media. The message arrives at the receiver who must decode the message. It is only when the encoding, transfer, and decoding process is effective will the message or information be successfully transferred.

Multinational team communications are more complex than homogeneous project teams as the team members have different first languages. The difference in first languages presents potential issues to the communication encoding and decoding process. That is when communications occur in different languages or in languages which are not part of encoder's or decoder's first or native languages the probability of communication errors increases. Add the probability of communication errors due to language differences to hundreds of potential communication paths and it quickly becomes obvious why multinational team communications can become very complex. Multinational culture research clearly identifies that communications between disparate national project teams is a prime source of cultural issues since culture is the foundation of communication.

Of the three communication system parts (encode, transmit, decode) the lack of shared language becomes the source of potential cultural issues. When one considers that there are approximately 2,800 different languages in the world and that each of these languages has a specific cultural underpinning it starts to become apparent how multinational project teams can experience major communication issues. It has been identified that communication is the fifth most important issue within the multinational construction industry (Lim and Alum 1995, 52).

The literature further articulates that multinational team communication variables include:

1. Attitudes
2. Social organization
3. Thought patterns
4. Roles
5. Nonverbal behavior
6. Language (Loosemore and Lee 2002, 518)

This list identifies the many cultural aspects underlying communications in general. To understand how each of these contributes to multinational team effectiveness and efficiencies, we need to expand on the context of how each can contribute or restrict the project team's culture.

A person's attitude is different than a person's value. An attitude "… refers to an organization of several beliefs around a specific object or situation" (Rokeach 1973, 18). What this means is that the set of beliefs which establish each team member's attitude establishes a positive or negative attitude toward the object of interest of the situation he or she is interacting within (18). This set of beliefs is grounded in past experiences or perceptions which were developed through cultural interactions over a period of time.

Within the project team environment, each member's set of attitudes can predispose that person to view those of not the same culture or perceived status in a positive or negative light. The person's attitude inclines him or her to respond to various interactions in a specific way. As an example, researchers have identified that not everyone is comfortable with interacting and engaging people of different cultures. This inability to interact across cultural lines is often founded within an existing bias or prejudice that is their underlying attitude (Earley and Ang 2003, 294).

The social organization variable refers to the individual's primary social interaction group. This variable recognizes that each of us belongs to a broader social group than the project team or the organization we work within. One way to look at this is within the context of a peer group. Your peers are those whom you routinely and frequently engage with socially, those you live with, your religious group, and political groups, to name a few examples. The metaphorical view that your social organization is your peer group is carried further in that your social organization, as a peer group, utilizes peer pressure as a means of forcing you to conform to the organization's norms. If these norms are in conflict with the multinational project team culture, a struggle between the two norms will occur. In this struggle (and as resisting peer pressure is always a difficult thing to do), the outcome may be that the peer norm will overcome the project norm. This event will directly influence the overall team culture.

In the area of the thought pattern variable, to be a little redundant, this is how we think while communicating. It is inclusive of the spectrum which spans from initial concepts, to reasoning, and problem solving (Xiuyan 2012, 54). Thought patterns are universal in the concept that all individuals have thought patterns. It is also universal that each culture has developed its own thought pattern process.

As an example, some cultures foster a very direct communication, *I* centered, thought pattern process where all information exchanges are conducted in a direct manner with little to no ambiguity. Some literature sources place the U.S. thought pattern culture within the *I* centered realm. This direct approach is viewed as blunt, abrasive, or rude to a culture whose thought pattern norms involve a more indirect approach which stresses saving face, providing an integrated group view, and one of stressing historical context versus predominate logical thought. The literature tends to place nations such as Japan within the indirect approach thought pattern realm.

When people with different thought patterns interact, a clash of cultures may arise. The clash in thought patterns can be highlighted within the example of a pure logical, *I* centered, discussion occurring with the indirect cultural view. In this situation, the *I* centered conversation stresses a pure logical argument while the other side is perplexed as the discussion is not taking into consideration the more subjective discussion points or historical precedence, which carry even greater importance to them. An example of different thought patterns is where the Western thought pattern is described as linear and logical based while the Chinese persist with circular, subjective, and group-oriented thought patterns. Obviously, it will be difficult to build a cohesive team if part of the team views some of the participants as rude, abrasive, and only willing to consider facts versus the group needs, personal relationships, and saving face.

The next variable considered is that of roles. Roles, as the word implies, is what we see as normal and acceptable behavior by various individuals in various positions. In some cultures, it is not acceptable for a woman to hold an authoritarian role or even to be employed outside the home. In other societies, there is no gender-based division of labor. The individual's skill sets, knowledge, capabilities, and desires are the driving forces as to where he or she will work. Our culturally accepted role view is based on what we have learned over time as socially acceptable behavior based on factors such as gender and social position. When a society's accepted roles run into conflict with the multinational team roles, it can create an environment which is totally dysfunctional and prevents interaction within and between the team members. A person with a deeply held role view that women should not be in the workforce will have a difficult to impossible time accepting a woman as a team member let alone as a superior.

Another potential multinational project team issue involves the nonverbal behavior variable. Nonverbal behavior is often described within the context of societies with high or low contextual cultures. By description, in high context culture settings nonverbal communications are extremely important. These settings and cultures are very selective in the words used, the setting in which the communication occurs, and the use of body language. It is these nonverbal clues which are the most important in how the communication is interpreted. In a high context culture, a contract may exist between two entities but it is the individual relationships and settings in which the communication occurs which shapes the interaction, rather than an explicit interpretation of the contract wording.

Conversely, low context cultures predominately rely on the literal verbal or written communications versus a heavy reliance on nonverbal signs. Low context cultures require communications to be very specific and explicit. Reading between the lines is not something that comes naturally to a low context cultural person. Low context cultures accept the written contract as exactly how the parties will interact.

In a multinational project team setting, a high context culture based individual will be confused as to why a low context culture individual is not getting the message as they were very explicit, in their view, in the details exchanged during the communication. At the same time, the low context individual can become really confused as to how the high context individual interpreted the conversation since the high context individual read more into the discussion than what the low context person ever said. One is reading more into the exchange than the other as he or she is heavily relying on the nonverbal clues while the other is not picking up on these clues but relies on the explicit message.

Delving a little deeper into high and low context cultures, the literature identifies that many of the relation-based countries, such as Africa, South America, Asia, and the Middle East, are high context societies where relationships mean more than contracts. What is essential to communication in these countries is the context in which it occurs. A Japanese business executive inviting you to a personal dinner has greater business meaning than if it occurred in a low context society. In the high context society this can indicate that the invitee wants to build a more personal relationship which equates to a better business relationship.

A famous example of a high context exchange occurred between the then U.S. President Jimmy Carter and Prime Minister Begin. The setting was the concluding meeting of the overall unsuccessful Israel and Egypt peace talks. During this last meeting, President Carter presented Minister Begin, from a high context society, pictures of the three heads of state who participated in the negotiations. On each of the pictures President Carter had written Prime Minister Begin's grandchildren's names. This high context exchange spoke volumes to the prime minister. More than words, this talked about peace or the lack of peace and the impacts to future generations, such as his own grandchildren. The pictures went far beyond the words.

Low context cultures, such as North American and most of Western Europe rely upon facts, contracts, and explicit wording more so than the context surrounding the conversation. In a low context society, an invite to dinner is generally just an invite to enjoy dinner together. Rather than rely on individual relationships, people in the low context culture rely upon explicit contractual wording and rules. Low context cultures view contract negotiations as predominately occurring within the formal contract negotiation session. This is different than high context societies where the formal contract negotiation session is viewed as the ceremonial signing session of an already agreed to contract. The actual contract negotiations and agreements occurred outside of the formal setting.

While nonverbal communication can be very important, overall communication comes down to the language used. If we could identify two identical nations that shared all norms, values, and attitudes, effective communications would still be very difficult if these nations did not speak the same language. As noted earlier, communications is a process of encoding, transmitting, and decoding. Adding the complexity of different first languages adds to the overall communication exchange difficulty. Even if one side of the communication exchange learns the other's language, communication issues can still occur. This communication barrier is due to the fact that that one side is a native speaker of the language and the other is not. What this means is that unless both sides of the conversation have immersed themselves in a common language culture, it is easy to miss the tonal inflections which alter the meaning of a word, what different length pauses mean, as well as how similar words may appear to mean the same thing but convey very different meanings.

The variations of communication are culturally driven. As an example, if a parent responds to his or her teenager's inquiry, "I'll think about that," what the teenager's hears is "no." The teenager is taking into account the contextual setting and interpreting his or her parent's response beyond the spoken words. This is a high context situation.

This is similar to a communication exchange which may occur between U.S. and Japanese business executives. In this exchange, being from a low context nation, the U.S. business person feels the need to be very explicit in his or her communication to ensure a full understanding of the contract language is conveyed. As such, he or she goes to great lengths in describing the details and providing a lot of information. The U.S. business person may even present examples of the various topics to ensure the true meanings are communicated and a full understanding is achieved. At the end of this exchange, the Japanese business person responds with "that is very interesting." In all probability, the Japanese business person does not find the dialogue interesting. In reality, they are just responding in a *face saving* manner. To the Japanese business person, as a low context person, his or her relationship with the other person means more than all the technical details.

The takeaway from this section is that multinational project teams face an even greater communication challenge than the homogeneous national culture project team. While ineffective and inefficient communications are faced by all projects adding to the cultural complexities of attitudes, social organization, thought patterns, roles, nonverbal behavior, and different languages extend the multinational project team's cultural challenges.

SUMMARY

This chapter presented discussions on project management organizations, sources of culture information as provided by professional organizations, an identification of some project management literature sources, and a brief review of relevant literature as it applies to culture research. This review included a look at homogeneous national project team as well as multinational project team culture research. The literature review identified that:

1. Project management cultural research is a very recent research area in the social science and project management discipline fields.
2. That, as with other areas of cultural research, performing project management based culture research is difficult.
3. The cultural dynamics of a multinational project team is similar but an expanded set of a homogeneous national project team.
4. The dynamics of project team culture are complex and can be described through system theory, complexity theory, and chaos theory.
5. The level of project management culture research has been expanding which is in alignment with the general management cultural research.
6. The project management system's cultural influences are an essential element for project team success as well as a potential impediment to project success.
7. There are many types of culture which include national, organizational, and subgroup or teams.

In summary, the literature identifies that teams can have a unique culture which diverges from the organizational or national culture. This culture influences the overall team success and its complexity. The literature also clearly identifies that the team manager or leader can and should foster a team culture which helps to establish a fully functioning, effective, and efficient organization.

Finally, the literature also identifies that culture is a complex system which is inherently a nonlinear, dynamic open system including a complex communication network.

CHAPTER 4

PROJECT TEAMS: THE SUBCULTURE WITHIN THE ORGANIZATION

All objects, all phases of culture are alive. They have voices. They speak
of their history and interrelatedness. And they are all talking at once!
—Camille Paglia (2013)

4.1   INTRODUCTION

This chapter explores the many facets of national culture, organizational culture, and subcultures. As an example, project teams or other work teams may have subcultures which are different than the organization as a whole which is unique from the national culture. All are specific yet all are contained within the systems environment. Within this complex cultural environment this chapter explores how these various cultures interact with, influence, and impact each other.

The primary objectives of this chapter are to (a) introduce the reader to the range of cultures which exist within the business environment, (b) present the reader information and discuss how various cultures (and subcultures) interact and influence each other, and (c) how different power levels, that is, formal and referential power as well as organizational structural power levels, have a moderating influence between the cultures within the organization.

Developing a greater understanding of how the various cultures are impacted and influenced by various power structures is structured within the open system theory research. Utilization of the open system theory provides a means of identifying and understanding social and organizational relationships which exist within open system business environments.

The chapter will also briefly discuss the issues associated with project subcultures as a lead-in to Chapter 5.

To begin the discussion on the types of cultures which exist within the organizational environment and their interactions, the chapter starts with a review of the unique cultural contexts which encompass national, organizational, and various subcultures. As highlighted in this book and found throughout the literature, culture is studied, analyzed, and researched within these categories. Yet, the research is not in full alignment with regard to the differences, similarities, and uniqueness between them. To partially address this and to develop a consistent theme for this book, the following sections expand the cultural classification discussion and descriptions.

4.2 NATIONAL CULTURE

Chapter 2 introduced national culture as the macrolevel culture which overarches all other cultures. National culture is the result of the nation's citizens' shared experiences, issues, and communications, National culture, as a macroculture encompasses all organizational cultures and the organization's subcultures which reside under or within the realm of the national culture. It is this overarching influence which impacts the organizational culture and the various organizational subcultures. While organizations have their own personalities and every person remains an individual, it is still a fact that each is influenced and impacted by national culture. Later in this chapter, these influences and impacts are presented within the open system theory.

4.3 ORGANIZATIONS: WHAT IS AN ORGANIZATION?

Moving down a cultural layer, from the national level, this section provides a description and definition of what an organization is, as well as, a principal view of its structure and culture influencers. This approach and discussion provide a structure from which deliberations on group and team cultures, that is, subcultures within the organization, can occur.

At the metalevel, organizations can be described as the structure in which for-profit, nonprofit, and government activities occur. The organization is inclusive of human resources, role structures, and social interactions which are intended to support the system's strategic plan. Stated in a slightly different way, organizations consist of a set of individuals working together to achieve an objective such as delivering a profit to the organization and its shareholders, meeting the nonprofit's objectives, or delivering the social needs which the government is best structured to provide.

To achieve any of these functions requires a multiplicity of functions, such as human resources, distinct roles and responsibilities, strategic plans, and group and team interactions. By description, human resources are the full set of individuals who are directly associated and work within the organization. These individuals jointly and independently perform functions and roles which support one or more of the strategic plan's objectives. While these individuals may have very divergent roles and responsibilities, collectively, they form an entity which is striving to achieve a common objective, set of objectives, or outcomes which are specific to the organization within which they are working.

To bring structure to the organization's human resources, various sets of individuals are grouped according to common functions within departments, groups, or teams.

Departments are sets of individuals who share a specialized functional area within the organization. Most major organizations will have an accounting or finance department, a human resource department, and a marketing department. Depending on the organization, there may also be an engineering department as well as an operations department.

Another aspect of a department is that all personnel will have the same employee–supervisor reporting structure. Further, these individuals also share and support the common set of roles, responsibilities, and department objectives. Even though a department is a very distinct entity, it can and often is broken down into smaller sets of individuals which are described as groups or teams.

When it comes to the terms, groups and teams, within the literature, these are often used interchangeably and imply a common meaning. From a common language or general discussion perspective this often works and minimal confusion occurs. People commonly refer to a team as a group and a group as a team. Yet, especially when discussing culture, there is a distinct difference between the definitions of a group and of a team.

Groups is a broader classification term in that any collection of individuals can be classified as a group. The engineering department may consist of several engineering groups or it may be a single group. Examples of various engineering groups include the woman engineering group, the civil engineering group, and the mechanical engineering group, and so forth. As the group classification examples indicate, each is inclusive of a set of people who have a common identifier which everyone relates to.

Further, groups are entities which collectively achieve a specific desired objective or set of objectives. Often the obtainment of these objectives is a result of peer pressure and common desires rather than a planned, scheduled, and detailed coordinated effort. The success of a group is generally measured by the result, not how they obtained the result.

A volunteer food bank group is an example of how peer pressure and group identification derive the end result. Within this group are individuals who share a common objective of helping to feed the hungry. They also share a common identifier as ones who are concerned for those who may not have the means to obtain sufficient food to avoid going hungry. While the group will probably have a leader, all members of the group are volunteers. To achieve the end objective requires the group to work together yet no formal power or command and control structure exists. Collaboration, cooperation, peer pressure, and a common vision form the group structure which allows it to meet its objective. Providing food for the hungry is the end result definition of success.

Rather than defining success by the end result alone, another method is to define success along the lines of the process as well as the final outcome. Defining success in this manner is applied to teams rather than groups. Teams are described as a gathering of individuals based on a complementary set of skills that are required to achieve a specific objective or a set of objectives. The set of required skills, within the team, is complementary and supportive as well as focused on the process and the final objective.

Project teams are classic examples of a team environment. Projects, and therefore their teams, focus on the process as well as the final objective. These are mutually inclusive in the definition of success. The team must be successful at the process as well as in delivering the final objectives. Failing to deliver the objective while performing the process well is a failure as is delivering the final objective yet failing in the process of achieving this.

An example of this is where the project team delivers the agreed to objective yet the end objective costs more than was agreed to, took longer than was agreed to, or failed to meet all specified quality requirements. By definition, this project is not a success. The process failed even if the end objective was delivered.

Other common organizational teams would include a safety team and an accident investigation team. Each team includes a set of individuals with complementary skills, a common process, and a common objective.

The formal alignment and grouping of departments, groups, and teams establish the formal organizational structure. It is easily defined and visualized by the organizational chart as well as the various job classifications' positional roles and responsibilities as well as job descriptions. The process and alignment of departments, groups, teams, and classification of individual positional roles and responsibilities define the formal, versus informal, power.

The formal roles and responsibilities' positional power structure identifies who can authorize what, commit the firm to specific financial obligations, perform what specific work, and who reports to whom. It also defines who can direct others to perform specific work as well as to apply discipline if it is required.

While all organizations must have a formal power structure, there is always a different power structure at work. This other power structure is the informal power or informal role structure.

An organization's informal role structure is a function of other power sources to include the ideas of expertise, relationships that one builds, respect earned from coworkers, and political power.

Expertise power is an informal power which is associated with individuals who are recognized as experts in their field. These individuals may not have company-granted positional power but they directly influence events, processes, and decisions due to their acknowledged and *accepted* level of expertise. Expertise power is often very visible. One can often quickly identify those with expertise power as one sees virtually everyone deferring to the *expert* whenever the discussion turns to their realm of expertise. Even if the *expert* is not within the room or meeting, it will be observed that someone will defer to or recommend that *Sally* or *Bob* be consulted as to which is the best way to proceed. This anoints them as the expert with a large informal power source whose input must be obtained before a final decision is made or a path forward is agreed to.

Another source of informal power comes from relationships that have been built over time. Sometimes this is referred to as referent power. Individuals with a high relationship or referent power utilize their communication networks, built on relationships, to *spread the word*. This informal power communication exchange can provide support for or opposition to organizational changes or leadership decisions. High relationship power members carry a lot of clout within the organization's structure. Those with relationship power may not be the technical expert but their ability to spread the word as they see it impacts the organization. This influence and power can be positive or negative. If the high relationship power source supports the organization's formal position their influence can be a positive impact. Conversely, if they disagree with the *formal* way forward their power can become an obstacle which the formal structure will have difficulty in overcoming.

Another informal power structure which is closely aligned with relationship power is informal power obtained from earned respect granted by coworkers. This is also an informal power that is obtained over time and through consistency in working with your peers. During these peer interactions, your coworkers have come to accept that you will perform as expected and probably go beyond the bare minimum. Your peers know they can count on you to deliver and excel at what is needed.

This performance consistency provides a strong informal power base. As your peers know how you will perform under most situations they grant you more informal power than others. Your word means more than others as you always deliver. This peer granted informal power is based on respect for your skills, capabilities, knowledge, and so forth. It is referenced as *respect informal power*. As an example, if in a discussion on how long an activity will take, if the one with a high level of peer granted respect informal power disagrees with the suggested activity duration his or her input carries a significant weight. That is, those peers who have granted them respect informal power take the input with greater validity than someone else's. What they are identifying is that the one with respect informal power always delivers and knows what it means to make a commitment. Therefore his or her input carries more weight than formal power does.

There are other individuals who are granted informal power not because they are recognized experts in their field, or have an extensive organizational relationship base, rather they have developed an extensive political acumen and know *how* to get things done. They understand complex social and business interactions and are able to influence what occurs without having positional power or expertise power. They have the political acumen and skills which are used to create opportunities to alter the current work path.

An example of this is the person who knows how to link the needs of various teams or individuals to meet them while ultimately obtaining what they need or want. A 1960s example is Radar in the television sitcom *Mash*. Radar knew whom to contact, what to exchange, and how to make things come together to achieve the desired end. Radar can be viewed as one with a high level of political acumen.

At the metalevel, the preceding discussion provides a very high-level view of what an organization is. That is, it consists of people working within a specific structure toward a common objective which supports the strategic goals of the organizational entity. It contains a formal power structure as well as a much broader informal power structure.

The preceding discussion serves this book's intent of providing a metalevel description and definition of an organization's structure, how personnel within the organization are grouped, and what formal and informal power sources exist. The reader is advised that organizational literature has numerous studies which provide a more detailed description and definition level of what an organization is and how the various power forms interact and can be leveraged. As an example, from one organizational literature research paper the reader can find a detailed discussion on "… contemporary organizational classification in the context of empirical, theoretical, and evolutionary perspectives" (Rich 1992, 758). This article and other papers in organizational literature note the study of organizations and that how to classify them is very complex with varying approaches, methods, and concepts.

Part of what complicates refinement of an organizational definition and its overall complexity is the human resource or social system. As with many other systems, once the human element is added to it, system complexity increases dramatically. Other factors, which increase the challenges of defining what an organization is, include developing an understanding of how the various subsystem interrelationships, such as administration, physical and responsibility structures, information, decision making, and economic and technical aspects interact (Hersey and Blanchard 1982) with each other not only at the process level but culturally as well. All these factors create unique organizational complex systems.

One approach to understanding the organizational complex system is through developing and better understanding the subgroups; such as teams, team networks, and groups, which make up the complex system and how these subgroup cultures interact with each other. This approach is discussed in the next section.

4.4   TEAMS, TEAM NETWORKS, AND GROUPS: IS THERE A DIFFERENCE?

The objective of this section is to present the reader (a) a set of definitions for various types of teams, (b) a discussion on the contextual structures for understanding project teams, (c) a discussion on why the project team culture is often referred to as a subculture, (d) a discussion on what differentiates project team culture from organizational culture, and (e) insight into the broad range of team research.

To achieve this section's objectives, the following presents, analyzes, defines, and highlights how literature differentiates between various teams' structural types such as team, team network, and groups. To frame this discussion, the following paragraphs are intended to provide an answer to the question of what and how a team is defined.

4.4.1 TEAMS

At the surface layer and in reviewing the extensive set of team literature, initially, a consistent theme and accepted definition are presented. The literature bounds the description of a team as a set of individuals who are focused or organized to achieve a joint objective or purpose. Each member of the team has a role and responsibility in supporting and aiding the other members in obtaining or delivering the intended result. These roles and responsibilities may be explicitly identified or implicit in nature. The roles and responsibilities are complementary where the combined team's capabilities and joint efforts are focused on their process and objective. Teams may be defined as virtual or physically collocated.

Teams are also found throughout the organization. The human resource organization may have a team or a set of teams. The accounts payable as well as the information technology department may have one or more teams. While the organization may have an engineering group or department, they may also have one or more engineering teams depending on what areas of responsibility they have, what the combined team process is as well as the team objective. Teams are also found in organizations that perform projects. Depending on the number of projects which are available, the organization will have one or more project teams.

The key factors to identification of a team is that the team will (a) have two or more people working together, (b) to achieve a common objective, (c) who are following a specific process, and who are (d) predominately stand-alone entities. That is, while the team is part of the organization, their focus, process, and objective is unique to them. Project teams are generally an ideal example of a team as they include more than one person with a common process and objective that are very unique to them. The project team is part of the organization, as a whole, but they tend to stand apart with their own culture.

The team, such as the project team, can be structured in many different forms and takes on a variety of responsibilities. By way of an example, a team's physical structure can take any one of the forms shown in Exhibit 4.1.

Exhibit 4.1. Team physical structure matrix.

	Dedicated	**Matrix**
Virtual	VD	VM
Collocated	CD	CM

VD: virtual dedicated; VM: virtual matrix; CD: collocated dedicated; CM: collocated matrix.

Exhibit 4.1 identifies that a team may be virtual or collocated. A virtual team is where the team members predominately or never interact in a face-to-face manner. All interactions occur using various technology sources such as teleconferencing, phone calls, e-mails, document-sharing software, and instant messaging. It is not uncommon for members of a virtual team to never shake hands or sit within the same conference room.

A collocated team, for this book, is the converse of a virtual team. The individuals on this team will predominately interact in face-to-face meetings, have actually shaken hands with each other, and physically see each other frequently if not on a daily basis. While a collocated team leverages technology communications, this set of people has the ready ability to "walk down the hall" to discuss something. They can and do hold informal, that is, water cooler, meetings where everyone is in the same room either formally or informally.

Membership in the team, as shown in Exhibit 4.1, can also be in the form of a dedicated or a matrix team member. A dedicated team member is someone who is assigned to and works within the team on a full-time basis. The direct reporting structure for the member is through the team leadership such as the project manager. While dedicated team members may normally be assigned to a different group or department, within the organization, such as the engineering department, for the duration of the team assignment, their primary reporting responsibility resides with the team and their assigned job and work activities are within the team. For all practical purposes, the dedicated team members' former organization remains a location that they will transition to at the end of the project, but during the project they have no functional responsibilities to that group, team, or department.

Matrix-assigned team members, for this discussion, are individuals who are assigned to the team, such as a project team on a part-time basis. As matrix-assigned team members, they have dual responsibilities, dual functions, and a dual reporting structure. One of their dual responsibilities is their *normal* day-to-day job which they were hired to do; or stated another way, their role within the functional organizational assigned group, team, or department continues to be a job requirement. As an example, if your normal job is an information technology network engineer you will still be supporting the organization's information technology department as a network engineer even while working with the project team.

The other part of the dual reporting responsibility is to support the project team you have been assigned to. Continuing with the network engineer example, his or her second responsibility would be assisting the project team, defining, engineering, and testing a network required to support a new technology or infrastructure while at the same time he or she will be responsible for engineering, maintaining, and repairing the current information technology infrastructure to keep things working within the normal day-to-day operating mode.

To delineate the matrix-assigned project team member further, consider an operations electrical engineer who works within the organization's engineering department. In this role, his or her normal job is to support operations electrical engineering needs to keep the manufacturing system operating normally and running within the decision constraints.

At the same time, this individual has been assigned to work with a project team. In this new assignment, the engineer's assigned role is to design, engineer, test, and commission a new manufacturing facility. At this point, he or she has a dual role of supporting the day-in and day-out operating facility while being part of a matrix project team which is designing, developing, deploying, testing, and certifying a new facility.

As a matrix project team assigned individual, he or she ends up with a dual leadership reporting requirement. This duality includes continuing to report to and take input and directions from his or her *normal* or functional boss. At the same time, he or she will be receiving directions from the project manager as well. This individual has entered the realm of two bosses which can and often does result in conflicts and issues.

The dual reporting relationship, strength, and potential conflicts are associated with the functional needs of the operating department as well as the project team needs. The duality of work requirements can and often results in conflicts in time requirements and workloads. What this means is that at any given point in time the day-to-day job requirements or project team job requirements may have different levels of criticality. As an example, if a sudden operational system failure is threatening to shut the manufacturing operations down then operations has a greater need than the project team. Conversely, if operations is running smooth and the broader project team is on hold until the matrix individual completes his or her assigned tasks then the project team has a greater need.

While supporting both operations and the project team, the matrix-assigned individual ends up walking a tightrope between the functional organization and project team roles and responsibilities. A significant contributor to how this effort transpires is a result of the functional department leader and the project manager strengths.

The leadership strength factor can be derived from a combination of formal and informal power structures. When competing needs arise, the leader with the greatest set of formal and informal power or authority will obtain his or her desired or required needs over the less powerful leader.

Formal power is the authority recognized and granted by the company. You find this power or authority represented in the organization's organizational chart, financial authority guide, and job description. These documents clearly establish roles and responsibilities, what decisions each member of the functional group or team can make, as well as who has the final decision-making authorization. Often matrix team formal power has greater complexity than regular functional organizations as the project manager often reports to a different senior executive than the various project team members' reporting structure.

Conversely, informal power, as noted previously, refers to aspects such as technical expertise, relationship, political acumen, and human relation skills that each individual has. As previously discussed and found within most organizations are people who are recognized as the expert in their technical field. As the recognized technical expert, final decisions and directions on how things are applied are often provided by these experts. Informal authority can also occur when the individual has advanced human resource skills which provide him or her the means to "make things" happen even though he or she has no formal authority. It is this combination of formal and functional authority which results in the relative strength of and sources of collaboration or conflict which can occur with matrix-assigned team members.

To recap, the attributes which identify or define a matrix, versus dedicated, team member are a matrix-assigned person (a) continues to work his or her functional organization role as well as his or her assignment on the project team, and (b) he or she reports to two bosses.

Referring back to Exhibit 4.1, a team can take any one of the following forms:

1. VD—virtual dedicated
2. VM—virtual matrix
3. CD—collocated dedicated
4. CM—collocated matrix

Based on the previous discussion, this book describes a team as a set of individuals who are assigned to work together, who have complementary skills, and have a common focus to deliver or produce a specific objective. The team may be a project team, an engineering operation's team assigned to improve a specific process, information technology team whose assigned objective is the continuous technology improvement process, as well as an evergreen quality team whose common objective is to ensure that quality continues to improve. Teams are found throughout the organization in many forms, features, and functions but share the commonality of more than one person who are working together toward achieving a common goal or objective while using a specific process.

Putting this into a project management discipline focus, a project team is characterized as (a) having more than one person assigned or involved, (b) who are working together toward a common objective, which in the case of a project is the agreed to scope of work, and (c) they are following a well-defined project implementation process.

Project teams are different from other teams in that they have a defined beginning and end. Other organizational teams, such as a quality improvement team, may have a definitive start but there is no defined end date. That is, the quality team continues as an operational team for an indefinite time period.

This set of characteristics is in alignment with the most often quoted definition of a team: "… a small number of people with complementary skills who are committed to a common purpose, performance goals, and approach for which they are mutually accountable" (Katzenbach and Smith 1993). Effective organizational teams will have all of these characteristics plus a common team culture.

4.4.2 TEAM NETWORKS

Another term found in the team literature is *team network*. By description, team networks differ from a project team or other organizational teams in how they utilize and rely upon individuals external to the team itself. That is, there is a core team which contains a specific set of skills and competencies which are insufficient to meet all the team needs. As an example, the scope of the project requires an ultradeep-sea drilling engineer's expertise. As a very specialized skill, the organization does not have anyone on the staff who normally fills this position. Therefore, there is a gap between the skills required to implement the project and what is available within the project team.

To fill the gaps, the core project team leverages their personal and corporate networks to identify and utilize specific personnel and their capabilities on an as needed basis. In the preceding example, the project team would collaborate with knowledge sources both internal and external to the organization to identify personnel who could fill the ultradeep-sea drilling engineer expertise skill gap.

A function of the team network is the purposeful actions of seeking and leveraging resources external to the core project team. These resources can and do come from internal to the organization as well as external sources.

The essential attributes of a team network are that (a) the core team does not have the skill sets or tools to perform the required task, so they (b) tap into the broader range of resources to (c) fill the short time span, activity-specific needs which ultimately achieves the team's goals.

Within this environment, personal networks and knowledge sources become invaluable (Cummings and Pletcher 2011). While team networks specifically and purposefully leverage external to the team resources, they still meet the core definition of a team which is a set of individuals who are working together to achieve a common objective.

The distinguishing characteristic of a team network, which involves looking external to the team for additional expertise and support, is sometimes referred to in the literature as team boundary spanning or team boundary management. As with team network, team boundary spanning leverages information, skills, tools, and techniques from experts outside of the core team (Marrone 2010, 912).

The application of team networking or team boundary management is increasing. A factor which contributes to an increase in utilization is the corresponding increasing number of predominately knowledge-based projects. Knowledge-based projects involve processes such as research and development efforts, software development projects, as well as highly technical systems development efforts. While a team may be formed which has all the technical skills and knowledge to deliver the final outcome, often, the core team lacks specific technical skills or unique knowledge. As noted earlier, to bridge this gap, the team looks externally to itself as a mechanism to address the missing knowledge need. As the literature notes, reaching beyond the core team occurs frequently when the activity or task is complex, or requires a very specific process, rare knowledge base, or technique. Teams also extend beyond themselves when the final outcome requires interaction across organizational boundary lines or even across organizations (Marrone 2010, 913).

An interesting aspect of team boundary management involves the interaction of the team's culture and the organization's culture. While later chapters explore this in further detail, the literature recognizes that culture influences the interaction and success of team boundary spanning. The team culture and the broader organization culture can either provide support for or hinder effective and efficient utilization of these external to the team resources. It is also possible that the external to the team resources may impact the team culture.

In summary, team networks are entities which meet the definition of a team. At the same time, they have a greater external focus than other team topologies. Team networks view the external world as a resource pool that—on a part-time basis—can provide those unique skills which will fill the knowledge gaps which exist within the core team.

In respect to culture within a team network, to increase the team's ability to be successful the team culture must include the essential elements of being willing to accepting input from these external resources and be willing to include these resources in key decisions.

4.4.3 GROUPS

Groups—What are they and how do they differ from teams? Earlier in this chapter, a general group description was provided. While the provided description is complete, it does not address or expand on many literature views one will find within the group literature. The following expands on how groups are referred to and described within the literature.

To start with and in alignment with this chapter's description, some authors take the position that there is a distinct difference between small groups and teams. A couple of examples include the 1986 research of Morgan, Glickman, Woodward, Blaiwes, and Salas and the 1993 research results of McGrath and Gruenfeld. While these sources are very specific on the differentiation between groups and teams, other authors fail to provide such explicit distinctions. That is, some literature sources apply the group and team terms interchangeably with no distinction or differentiation at all. These authors interchange the group and team terms throughout the paper, book, or article indicating that a group is equivalent to a team or a team is equivalent to a group.

The literature also presents two distinct views as to where the group resides in relationship to the organization. In one research stream, a group consists of individuals who are internal to the organization. As the whole group resides within the organization, all members of the group freely interact with each other and the broader organization. This interaction ensures a higher cognizance of the other participants' actions, effects, and subsequent consequences. That is, they know what the others are doing on an almost continuous basis (Heise 2013, 54). The organizational commonality provides a foundation for achievement of the group's objectives.

The alternative literature view is that a group is an exogenous entity where the membership and group boundary are explicitly defined as those external to the organization specifically (Katz et al. 2004, 311). As an exogenous entity, the group members are external to the organization and may not be fully aware of what is occurring within that organization, specifically. Aligned with this view is that when a set of individuals interact internal to the organization they form a team but external or exogenous entities are groups rather than a team.

The exogenous view blurs the clear distinction between groups and teams as specifically set forth earlier in this chapter. That is, rather than the definition of groups and teams being focused on the process, objectives, and interactions, the exogenous view is each small cluster of individuals working toward a common goal is a team regardless of how success is to be defined. The exogenous view is not supported by this book.

The literature also identifies a common concept that all groups exist within a physical environment, a social environment, and they exist within a larger cultural context. It is within this dynamic environment where the literature identifies how different cultures can and do exist within internal as well as exogenous groups, organization silos, teams, and the broader organization as a whole (Levine and Moreland 1990, 589).

In summary, this section provides a look into the characteristics and attributes of teams, network teams, and groups. While attribute variations exist, within the literature, between the various contexts, at the core level they all have some common attributes such as consisting of a set of individuals who interact, at some level, on a fairly routine basis. A group's interaction tends to be less frequent than a team and not as focused on defining success to include the process.

Project teams are specific examples of an organizational team which shares commonalities with other organizations' teams but has an additional characteristic, a very explicitly defined termination time, which other teams, in general, may not include. That is, in looking at a team's life cycle it includes a formation or beginning, a process time, and a termination.

What this means is that all teams have a beginning. An example of this is the safety team was established following an audit which identified an issue with the organization's safety culture. The new facility development project team's beginning is correlated with the organization's decision to develop the new facility and assigning the resources to the team to achieve this objective. In each example, the set of individuals passes through a phase where the team members come together to form an entity with a specific focus.

Immediately following formation, each of the teams then proceeds through a development phase or process where the team culture starts to emerge. The development process is a social integration process (Levine and Moreland 1990, 589) which results in a final entity with its own culture. Within this culture are the specific team's attributes of a status system or structure, norms, values, attitudes, problem resolution processes, and communications.

At this point, project teams and organizational teams diverge in that project teams have a termination date, the end of the project, explicitly defined in the beginning. At the conclusion of the project's joint objective, the project team disbands, the termination phase, and the team members move on to other work. Conversely, other operational teams continue to exist with no explicit termination point. While the organization safety team may eventually terminate, at the initial safety team formation an explicit termination date may not be established. Thus, project teams form, do their job, and then disband. Their life cycle is complete and defined from the beginning. Organizational teams, on the other hand, form, do their job, and continue on. They generally tend to have no specific end date which is known up front.

Thus, project teams are a specific type and a recognized subgroup from within the organization. Project teams may include people from different nations and organizations, yet they are keenly aware of the shared objective of delivering a specific outcome and the process to achieve this outcome. The project team defines success as inclusive of the process as well as obtaining the end objective. A project team's culture may predominately reflect the organization's culture or it can establish a culture which is significantly different than the parent organization with its own set of reporting requirements, assignment processes, and a unique set of operating procedures. The following sections expand on this concept further.

4.5 CULTURES AND SUBCULTURES

The phrases *culture* and *subculture* have distinct histories. As discussed earlier in this book, *culture* first appeared in the 15th century. From an organizational perspective, the culture phrase can be used to describe what an "… organization 'has' as compared with something an organization is …" (Sackmann 1992, 141). At the metalevel, discussing an organization's culture refers to the inclusive set of problem-solving techniques, shared values, acceptable interpersonal interactions, shared norms, and shared basic assumptions. Generally, the organizational culture literature discusses an organization's culture as if all members, groups, and teams share all cultural attributes at the same level and to the same degree.

Conversely, there is a body of organizational culture literature which challenges the concept of an all-inclusive adoption of a single culture which applies to the organization as a whole, all internal groups, and teams. That is, this body of literature discusses that more than one culture may exist within the organization. What this identifies is the concept that organizations have a culture as well as potential subcultures.

Within the view that organizations include different subcultures, the discussion begins by identifying the organization culture as what one would see by analyzing a very broad section of the members from across the organization as a whole. This observed organization culture is very similar to a nation's culture, in that the observed organizational culture will not explicitly define or describe the culture of each member of the organization. As national culture describes the generally observed culture of that nation's members, it does not and cannot be explicitly used to describe any specific individual's culture. Defining and discussing an organization's culture is to define and discuss the metalevel culture, not the individual's culture or the various subcultures which may reside within the organization.

The phrase *subculture* was not introduced into cultural research efforts until around 1986. Prior to this, culture research predominately focused on national or organizational universal cultures. The literature generally lumped everyone's culture as the culture of the nation in which they resided or the organization where they worked. In this context, subcultures or variations in national and organizational cultures were not taken into consideration. This view has been changing since research introduced the idea that subcultures existed within nations and organizations.

One way of defining an organizational subculture is it consists of a group or team or distinct subset of individuals working together who are distinct from the broader organization and they hold a common view that they are unique from the overarching body. These distinct entities further share a common problem set, decision-making process, and interacting norms which are different that the organization in and of itself (Bellou 2008, 499).

As is discussed further in later sections, national, organization, and subcultures can be described in several ways such as the various cultures are highly consistent within the higher or overarching entity, very divergent, or somewhere in between. Exhibit 4.2 depicts some possible consistent culture ranges as two different statistical curves. The leftmost curve, Highly Consistent, represents a team whose members have a very consistent set of cultural attributes with one another. In this context, there is a high level of agreement, acceptance, and application of a common culture between the members.

A very functional project team may be an example of such a highly consistent culture. In this environment, the various project team members know what acceptable values, norms, and communication styles are and how decisions are made. A cohesive and consistent culture reduces conflict and provides a firm foundation for a successful and efficient project team.

The right curve, in Exhibit 4.2, is intended to demonstrate a team whose members do not fully share a common culture or are less integrated in the general team's cultural attributes. Within this environment, the team members may hold other cultures' attributes, rather than accept the new team's culture. This can result in a team structure which fails to share values, norms, assumptions, attitudes, and decision-making processes. That is, the statistical cultural span is much wider than the homogeneous cultural world within the highly consistent structure. This typology is consistent if the group of individuals analyzed, or the entity of interest, is a group or a team. That is, the entity of interest may have a very integrated and aligned culture or a very diverse and may be even dysfunctional culture.

Exhibit 4.3 provides a view of potential relationships between an overarching culture and various subcultures which may exist with the environment. As this exhibit demonstrates, the level of culture assimilation and team integration may, and in all probability does, vary between various subsets. There may be culture overlap between the various sets of individuals which would result in a blending of cultures or there may be

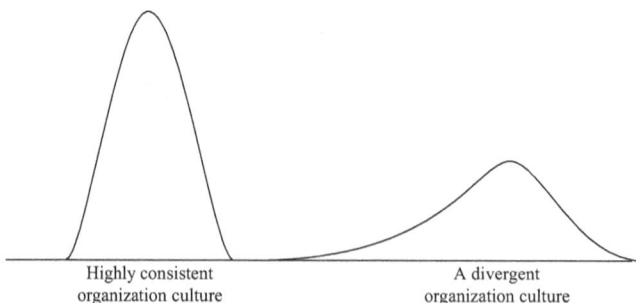

| Highly consistent | A divergent |
| organization culture | organization culture |

Exhibit 4.2. An example for organizational cultural ranges.

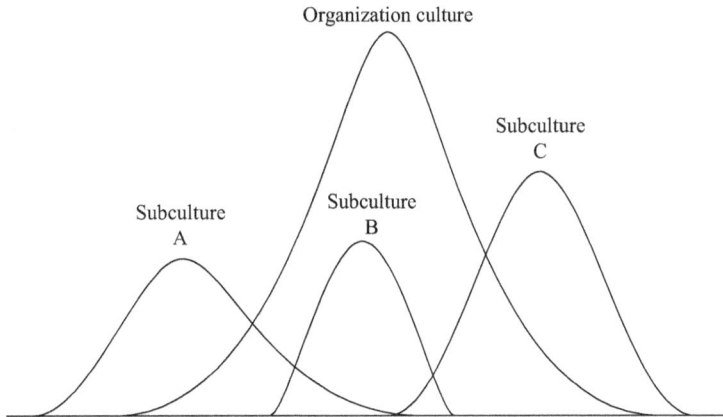

Exhibit 4.3. Range of organization cultures.

no interaction between the subcultures. The lack of interaction increases the probability that the divergent cultures will be very distinct.

Up to this point this chapter has discussed the literature constructs of organizations, groups, and teams. The intent is to clarify the differences and similarities between the various entities. So, at this point the reader should be very comfortable that when the discussion turns to culture one can present the discussion within the context of nations, organizations, and groups or teams. Yet the unanswered question so far is: Is there just one national culture and just one organizational culture at work within each context? The remainder of this chapter explores this question and presents a literature-based answer for the reader to consider in his or her own work environment, studies, and research.

4.6   NATIONAL CULTURE: IS THERE REALLY JUST ONE NATIONAL CULTURE WITHIN A DEFINED NATION?

So what is national culture and is there only a single national culture within a specific nation? Before digging too deep into this discussion and to help frame it first and foremost, we need to understand what a nation is. You cannot answer the question about there being a single culture or multiple cultures within a nation if the definition of a nation is not explicitly clear.

When discussing what a nation is, the literature is rife with inter-changeably applied terms such as *nation, country, nation state,* and *State.* Often these terms are used interchangeably, as if they all have the same definition. Yet, as the following paragraphs clearly identify, each of these terms has its specific definition and explicit association with culture.

To start with, the word nation is from the Latin term *natus*. *Natus's* Latin definition means *to be born* ("Latin Word Study Tool" 2013). It indicates a common lineage based on birth. In the past, all Irish were born of and could trace their lineage from a very specific Irish bloodline. This is the definitive understanding of *natus* or someone who was born to a specific bloodline. Today, someone born in Ireland may or may not be of 100 percent Irish descent. Someone born within Ireland may trace their ancestry through a German lineage that moved to Ireland several decades ago. While the current descendants may view and align their association with the Ireland nation, they are not explicitly Irish as defined by the *natus* term. The same could be said of the original Navaho Indian nation. To be of this nation requires you to be born of parents who could trace their blood lineage exclusively to Navaho bloodlines.

Over time, owing to the increased mobility of the world's population, intermingling of pure nation bloodlines, and common terminology application, today's meaning of a nation is no longer restricted to a pure bloodline descendant. Today, the definition of whom is included within a nation has been expanded beyond that of a common blood lineage to be inclusive of those who share a common "... postulated interrelationship—a 'blood' bond between members. This blood relationship may be actual, but more often, it derives from myth" (Rasmussen 2001). Nation is now inclusive of all those who share a common heritage rather than a pure bloodline relationship.

For a nation to exist, its members should share a common cultural heritage with all aspects of a shared culture background, such as its artifacts, values, and norms. The culture artifacts are a critical component as they "... represent the 'patrimony' of the nation, and is often invested with considerable sentimental values ... " (Rasmussen 2001). The nation's members also share a common language. While many languages may exist within the nation, there is a primary language that is specifically assigned to that nation as a definable attribute. As an example, those native to France speak and communicate in French. At the same time, those native to the United States speak and write U.S. English, not British English. Members of a German nation all share a common language or German. This is not to imply that no other language is spoken within a nation; just that from an internal and external view, the principal language spoken within a nation is that nation's primary language.

People who share a nation also hold an accepted and deeply held association with the nation by those who originated there. People, who either currently live within the nation or if they have even moved to some other geographic location, refer to themselves as members of that nation. While Germans reside within the United States and claim to be U.S. citizens they, and others, still refer to themselves by their German ancestry and national association. They, for all practical purposes, are of the German nation, not of the U.S. nation. Along this line, people refer to themselves as a second- or third-generation Greek family who maintain an indelible cultural link to a different nation than that in which they currently reside. The research shows that this association is deeply ingrained and virtually non-negotiable. My nation is my nation, irrespective of where I may currently be living.

Thus, a nation can be defined as a group of individuals who have a common postulated interrelationship which transcends factual blood relationships. A nation also has a shared common culture and a common language. The common language metric does not mean that all nation members speak only one language. What this means is that while different languages may be in use, a common, deeply ingrained, and closely held language will prevail and be clearly associated with the nation.

So, how is a nation different from a *State*? The term *State* originates from the Italian language, specifically from the term *lo stato*. *Lo stato* generally denotes an entity which holds power over those within its structure. This state power base provides those in charge the authority to command the members to act in a specific manner, define right from wrong, and punish or reward accordingly (Rubinstein 2004, 151). The literature expands on this description such that *lo stato* is inclusive of all social and political activities which result in the ultimate ruling and governing of the country. *Lo stato*, therein a state, is a social structure which involves entities with legal and legitimate authority over the resident population.

A state is different than a nation as a state does not require a common culture or language and or a strong member association. A state has institutions, structure, and authority of control over the population. States can and do include multiple nations within their realm of authority and control. As an example, Canada is a nation but it contains states that have specifically authorized institutions, structures, and authority of control. At the same time, the state authority of control extends to various nations which reside with Canada such as the First Persons nations.

The term nation-state is of recent origin where it was first mentioned somewhere between 1915 and 1920. As the hyphenated word, this term brings together or joins the definition of nation and state. Another example of the use of hyphenation is where two people join together in a legal framework, such as a marriage. An outcome of this merger is that rather than one party taking the name of the other they create a hyphenated name of both. This indicates that while they continue to be specific entities, they have elected to become one that has a new, combined, name.

Therefore a nation-state is an independent state inhabited by people of one nation and one nation only. Within this context, those who live within a nation-state share a common cultural heritage, language, and hold a deeply ingrained association. The nation-state also has a set of institutions, structure, and authority of control over the population. This is the state portion of the name, which now includes legal and legitimate authority within the nation as well.

A keen observer will have noticed that the definitions of nation, state, and nation-state have not included explicit reference to geographic borders. That is because these terms are not geographically defined. Rather, the definition of the *country* term explicitly includes geographic borders. The word country is of Latin origin which originates from the word *contra*. *Contra* is specific in its reference to a distinct area of land with clearly defined borders ("Country, Nation or State?" 2011). As such, when looking at the world today, one sees it divided up into many different countries as each country is geographically defined and borders between countries are explicit.

Unique to each country is the fact that at least one state exists within that country. All countries have a set of institutions, structure, and authority of control over the country's population.

When looking at the world map there are about 195 identified countries and associated states. Depending on world events the actual number will vary. Regardless of the actual world country count, the final number does not identify that there is an equivalent number of nations in the world. The people of a nation can and do cross geographic and political boundaries as well as state control. The people of a nation can reside within a single country and state or they can cross country and state geographic boundaries. That is, nations are not geographical border bound or subjected to the exclusive subject of a single state authority and control.

Exhibit 4.4 demonstrates the relationships between two different geographically defined countries and four different nations within these countries.

As shown in Exhibit 4.4, Nation 1 demonstrates or depicts the situation where a native or indigenous set of people, who constitute a nation, physically live in a geographic region which crosses a country's borders. This situation can be found where the U.S. and Canadian geographic borders exist. In these areas a common indigenous nation has existed for

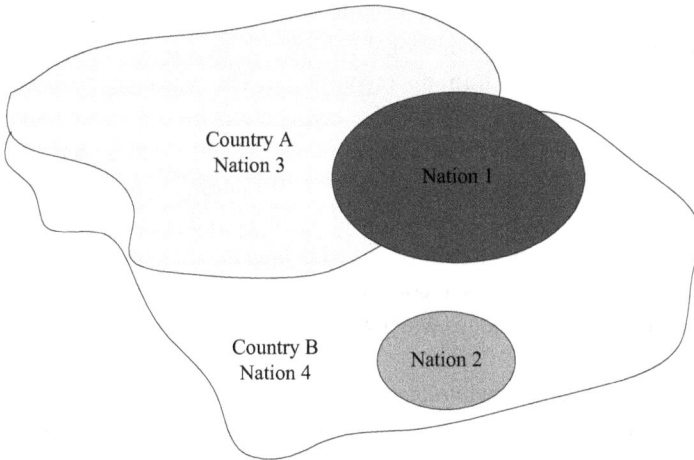

Exhibit 4.4. National to country boundary comparison.

generations. This lineage stretches back into the distance past with a shared culture irrespective of the country's geographic borders or state control (Starks, McCormack, and Cornell 2013, 1). These indigenous nations existed in the past and continue to exist as a single nation with a shared national culture irrespective of geographic borders.

In this example, the indigenous people's location and culture exist on both sides of the two states' geographic and political boundaries or boarders. Their culture continues regardless of geographic borders or state institutions. They are a nation of people within different countries and different states.

One key takeaway from this discussion is that when culture is discussed, at the top level, it explicitly refers to a nation which may or may not be fully associated with a country or a state. This sometimes results in confusion as an author may refer to a national culture in the context of a country or state when the nation may span geographic and political boundaries.

4.7 ORGANIZATION CULTURE: WHAT IS THE RELATIONSHIP WITH NATIONAL CULTURE?

The literature is clear that organizations have their own culture. Yet, organizations are contained within a nation or they even span different nations. As such, what is the interaction between national culture and organizational culture and how are they different?

The answer to this question begins with the description of an organizational culture. An organizational culture encompasses the interactions of a set of people who form the organization. This set of people, through a process of interacting, communicating, and making decisions, develop a core and shared set of basic assumptions, beliefs, values, and norms, which is a cohesive set of shared cultural attributes. The specific makeup of these attributes is what creates the uniqueness of that organization. This uniqueness is demonstrated or observed in how the members interact with each other and their external connections, how they make decisions, their observed rituals, and values which are applied internally and externally to the organizations.

The information technology company Apple is a prime example of what a unique corporate culture is and how that culture enhances the organization's bottom line. From numerous sources Apple's culture is defined as unique, one that cannot be duplicated, a culture where management pushes their employees extremely hard while focusing on secrecy and finite attention to detail. The key here is that this is the corporate culture which their employees, not only management, demonstrate internally and externally to the company. This is their demonstrated and observed artifacts, values, norms, and decision-making processes.

A new concept, which this chapter brings to the organizational culture discussion, is that organizations can and do exist within a single or homogeneous nation or they may also exist within multinational locations even if they reside within a single state. The multinational existence occurs when the organization's operations are collocated in different nations even if they reside within a single state. As an example, Turkey has significant nation populations of Kurds, Greeks, Jews, and Armenians. They all have unique and distinct national cultures which are not the same as the Turkish national culture. In this situation, if an organization had offices predominately in a Turkish part of the state as well as within the Armenian-dominated area, the national influences on the organization's culture may be different at each location. The differences would reflect the higher order national culture influences and impacts on the organization's culture.

How a nation's culture may or may not impact an organizational culture, within its boundaries, lacks clarity and concurrence among national cultural researchers. Organizational culture research is equally ambiguous as to the real impact or influence which a nation's culture may or may not have on an organization's culture within the nation's environment.

While full agreement is not identified, within the various literature sources, a common literature theme which does occur is that if one culture influences another it is from the most powerful to the less powerful culture setting. In this case, the national culture would influence and affect the organization's culture rather than an organization's culture being developed solely from within the organization. Stated in another way, cultural influences, if they occur, flow from the metalevel down. In this case, the cultural influence is from the national level toward the organizational culture, not from the organization to the national culture. As an example, the Global Leadership and Organizational Behavior Effectiveness (GLOBE) research study stresses that there is a strong relationship between national and organizational culture (House et al. 2004). This literature theme, when it occurs, is consistent in that a nation's culture directly influences, constrains, and impacts organizations' cultures, not the other way around (Johns 2006, 396).

The reader should also be aware that there is a counter or converse literature theme that the concept of national culture influencing an organization's culture is in conflict with management research. What this means is that within management research there is a stream of research and literature which identifies organizational culture as one which emanates from the organization's leadership only. It is the leadership that is key to establishing the organization's culture and sets the organization's direction. These cultural influences are only possible if the organization is not overly constrained or limited by influences at the national level (Gerhart 2008, 254). This information also indicates that if the leadership culture is not the same as the national culture, one can expect a significantly observed cultural difference between the two.

While the national and organizational cultural literature is not in full concurrence, one approach to visualizing the potential impacts national culture may have on an organization's culture is through an open system lens. As system science, physics, and the study of complexity have clearly shown, open and closed systems exist. The distinguishing difference between these two systems is associated with how the system interacts, or not, with the external environment. If the system interacts with the external environment, by accepting external inputs and outputting work, then it is an open system. A university setting is an open system example as it receives input from a vast array of sources external to itself while outputting many items, such as research, graduates, social events, to name a few, to the external environment. If there is no interaction with the external environment, it is a closed system. From a cultural perspective it is difficult to identify any closed system. Culture, as defined, is a continuously changing system based on external and internal events. These events also impact the outflow of energy from the system to the environment. While the transformation process may be slow, it is continuous and it impacts what occurs within the system as well as how the system interacts with its environment.

It is the aspect that an open system must interact with the external environment which provides an avenue for understanding how national culture can have an impact on the organizational culture. Exhibit 4.5 provides a view of how this open system interchange occurs.

As represented in Exhibit 4.5, the organization exists within, not external to a national culture. The people who work within the organization also live, work, and interact within the nation. The organization also interacts with the nation as it supplies services, material, or finished products to those within the nation as well as entities external to the nation. Further, the organization's interaction with the national culture occurs as the sets of people transitioning between the cultural environments as they move from work to play, to family life, so forth.

Thus, based on an open system view, national culture will have some level of impact and influence on an organization's culture. That is, those who move between the national and organizational cultural boundaries bring with them the external culture, to a certain degree. As noted further later in this chapter, there is greater influence from the national culture to the organizational culture. This is a reflection of a nation's greater power influence over any contained organization.

Within this context, the next culture-level question involves organization subcultures or per one posed question, "If [organization] culture is composed of subculture[s] where do they emerge and what triggers them to emerge?" (Sackmann 1992, 140).

The answer is that subcultures occur as most organizations comprise many smaller, differentiated, departments and groups, teams, or both. When one analyzes an organization, it is a collective set of departments

Exhibit 4.5. Open system culture influences view.

and groups, teams, or both with distinct roles, responsibilities, tools, techniques, and educational backgrounds. Each of these entities encounters a unique set of problems which must be solved, which may have different communication interaction requirements, and encounter settings which require different decisions. Examples of this would include the cost accounting team, the engineering department or team, project teams, and quality teams, as a few examples. Taken as a collective whole, the differentiation between internal groups and teams results in varying cultures or stated another way, they appear as the organization's subcultures.

As an example of how different teams may develop different subcultures, take the comparative case of an engineering team and a sales team within a single company. The personnel drawn to each of these entities generally have different personalities, education, and risk tolerances. Exhibit 4.6 demonstrates the different characteristics which are identified within the typical engineering and typical sales teams.

What Exhibit 4.6 demonstrates is that those who are typically drawn to different fields not only have different educational backgrounds and skill sets but their personalities and approaches to problem solving and decision making are generally different as well. In the end, each of these teams will build their unique subculture which is consistent with their team setting, interactions, decision-making requirements, and problems.

The next question is, does the organization's global culture impact and influence its subcultures and if so how? As discussed in the literature, the answer is yes, the organization's culture can have an impact on subcultures within it. As with the earlier discussion of national culture impacts on organizational culture, the organization and its subcultures form an

Exhibit 4.6. Example of culture traits.

Engineer	Sales
Heavy reliance on logic	Persuasion is a stronger skill set
Very detail oriented	Looks at the *big* picture
Very comfortable working alone	Interaction with others is required and desired
Strive for order and structure	Comfortable with greater ambiguity
Desire to minimize risk as much as possible	Willing to take greater risks
Often are perfectionists	Good enough is good enough, perfection is not required
Often characterized as an introvert	Often described as an extrovert

open system. In this situation, they share energy in many forms as they interact with each other. What this means is that as the subculture member interacts with the broader organization culture they may adopt part of it and transfer attributes of the organization culture within the subculture. This provides a modifying and altering impact to that subculture. While it is possible that the subculture may influence the organization's culture, due to size and complexity there is less reverse cultural attribute impacting flow; that is, there is less cultural impact from the subculture to the organization culture than from the organization culture to the subculture.

Why is it important to understand the various interactions between national culture, organizational culture, and subcultures? The answer is based on the foundation that culture drives how the group or team members interact, communicate, solve problems, and make decisions. Having an understanding of how these cultures interact is required to develop a deeper understanding of the various cultures and their interactions.

Therefore, subcultures are unique sets of groups, subgroups, or teams within the organization as a whole. As organization cultures may be different than the national culture, the organizational subcultures similarly reflect their members' interpretations of reality which may be different than the organization's culture as "Subcultures similarly shape their members interpretations and actions but often in divergent ways" (Howard-Grenville 2006, 47). Thus it is important to understand how the various cultures may or may not interact and how the power relationships drive specific interactions while excluding others.

Follow-on questions to the preceding paragraphs include: Do subcultures influence each other and are all subcultures of equal stature and power?

When analyzing an organization with multiple subcultures, a point of consideration is the interaction between these various entities. The potential interactions span the spectrum from no interaction to very intensive in between teams collaboration, interaction, shared problem resolution, and decision-making requirements. As the interactions cover a spectrum of potentials and in alignment with systems science, as each set of people extends and expands its direct dealings with another group or team, this results in a sharing of energy. This energy can be in the form of information, problem resolutions, communications, norms, values, and shared decision making. This sharing of information alters the information sources. Thus, the divergent subcultures, as the sharing information sources, will adapt to the new environment as well. On the other hand, if there is no direct interaction between groups or teams no culture modifications will occur. Cultural modifications occur through interaction, so when interaction does not occur the various cultures are not altered.

While there is logic behind the concept that divergent subcultures may influence and provide energy for culture changes in other groups or teams, one has to be aware that all subsets of people are not of equal political stature as well as positional or technical power authority. This is highlighted in the literature as "… subcultural groups are not equally powerful, with some enjoying status that flows from centrality in the work organization, or irreplaceability of their expertise or skills" (Howard-Grenville 2006, 51). In this situation, the lower power subculture may mimic or use the higher power level subculture's values, norms, and attitudes rather than its own. The application of this working environment power structure results in changes to the lower power authority team culture.

The discussion on organizational subcultures highlights the aspect that subcultures can be influenced by the overarching organization as well as individual subcultures have the ability to influence other subcultures, and that groups and teams with greater organizational power have greater influence on groups or teams with a lower power level and are usually not influenced in reverse. That is, the culture source with greater power has greater impact on less powerful subcultures. Conversely, lower power source cultures will not overly, if at all, impact higher power source cultures.

Exhibit 4.7 demonstrates the overall potential interactions and influences between different cultures. As noted earlier, national culture may

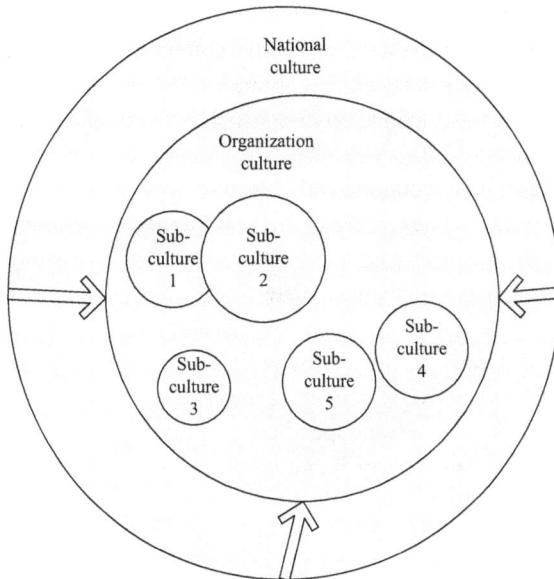

Exhibit 4.7. Overarching culture influences.

impact the organizational culture. At the same time, organizational culture may impact the group, team subcultures, or both. It is also possible that within organizations, multiple subcultures of various positional or technical power authorities may exist. Depending on the power relations and levels of interactions between these subcultures, the subculture can also influence and impact other subcultures.

At the same time, as shown by subculture 3 in Exhibit 4.7 an organization subculture may stand alone in relationship to other subcultures. This subculture may only be influenced by the broader organization culture, not their peers.

4.8 PROJECT TEAM CULTURE: WHAT IS THE RELATIONSHIP WITH OTHER CULTURES?

To establish the discussion on project team culture and its relationship to other cultures, the first step is to identify the environment in which the project team is operating. Specifically, one type of project team exists in an environment where:

1. The team members are on temporary loan to the project team.
2. The team members are assigned to the project full time.
3. The team members interact with other team members frequently, such as daily.
4. The team members are faced with a common set of problem resolutions, and decision-making opportunities.
5. The team has a defined process which is designed to obtain the end objective.
6. The team has a common end objective.
7. The team defines success both in terms of process and end objectives.
8. The team has a defined end date.

Within this context, this set of individuals becomes a project team with a common objective of delivering the project outcomes. But, will this team structure develop a team-specific subculture?

The project management literature presents a trifurcation answer to this question. On the one hand, the literature indicates that within a single organization project teams have a homogeneous culture. That is, all project teams' cultures will be nearly identical. This cultural commonality is derived from the organization's metalevel culture.

Other literature sources identify that within an organization the various project teams' cultures will be heterogeneous (Auch and Smyth 2010). The heterogeneous set of project teams' cultures will be significantly different. This heterogeneity is derived from each project team facing unique challenges, different problem sets, and different communication requirements.

Finally, there are literature sources that suggest the concept that the project team will consist of a set of people with their own cultures who are brought together for a common objective. This set of individuals will maintain their own culture versus establishing a project team culture. Such an example is the ad hoc project team.

Whether the project team is homogeneous, heterogeneous, ad hoc, or a combination of these cultures, understanding the various cultures and their potential impacts is increasingly important (Yazici 2011, 20). In the current world of ever increasing complex and fast-paced projects, ensuring a project team is efficient and effective is of paramount need. To this end, the project manager and the broader organizational management structure must understand and foster the most effective project team subculture which supports overall project success. The positive and negative impacts that culture has on the project team's success is well documented. A cohesive project team culture results in greater potential for project team success.

Developing an understanding of project team subculture requires an understanding of the project team culture trifurcation view and how the potential results reflect the dynamics of culture development within different contexts. What this means is that the project team culture is developed through shared interactions, resolving common problems, developing accepted interpersonal interactions and communications, developing decision-making processes, as well as establishing accepted values and norms. When one views an organization's project teams' inter- and intrapersonal relationships, interactions, common problem sets, values, and norms, a vastly different view may be present across all project teams. In one context, the organization's project teams may be very closely linked and overlap in many areas. They will share common experiences, needs, wants, and ultimate desires. This environment can foster a more homogeneous culture development and modification process. Culturally, the teams will appear to have similar heterogeneous team cultures rather than heterogeneous or ad hoc cultures.

Heterogeneous project teams, conversely, may be the result of various project teams working within significantly different environments where there is little to no overlap in work settings, or interpersonal interactions, problem resolution needs, or team decision-making processes. In this setting, each team can be viewed as an island within the overarching organization. This island is clearly separated and isolated from all other project teams. This situation fosters an atmosphere where each team will develop its own culture. The end result is a heterogeneous set of project teams' cultures within a common organization.

The third leg of the potential trifurcation occurs when the project team consists of a set of individuals who just do not form a common culture team. This lack of a common project team culture formation may occur for many reasons. On one hand, the lack of a cohesive team culture can be the result of a very short, intense, project where team member interaction virtually never happens. In this situation, every team member is focused on his or her own assignment, needs, wants, and desires with no real need to interact with the other team members. Each can be viewed as an island within the project environment separated from the other islands.

The lack of a common culture may also occur if it is a dysfunctional project team with conflicting and competing individual objectives, desires, and needs. The resulting internal conflicts hinder or prevent the development of a shared culture.

A specific project team structure which hinders development of a shared culture is the ad hoc project team. "The Ad-hoc team is a group of people who work together but have complex reporting lines and allegiances outside the basic team group and who are members of multiple teams and groups" (Handy 1989, 1). In short, the ad hoc project team is characterized by the view that each member is first and foremost looking out for what is best for him or her, not what is best for the project. This context limits or even eliminates the potential of developing a common team subculture as the team members are not sharing or interacting toward a common goal but working toward an end which enhances their own, rather than the team's, position.

While there may be a range of project team cultures, within an organization, the team's culture is still a subculture within the organization. This statement is based on the context that a project team consists of a collective set of individuals with a distinct and common set of roles, responsibilities, tools, techniques, problems, and decision-making requirements. The project team may also have unique communication interaction requirements as well as inter- and intrapersonal interaction requirements. This environment, which exists within the broader organizational environment, becomes a subculture.

Within the general definition of a subculture, as was demonstrated in Exhibit 4.7, the project team's culture is part of the organization's open system environment. It receives and transmits energy between itself, the organization, and potentially other subcultures. The interactions of these groups and teams will have different impacts on each other. The organization, due to positional authority and structure, has a greater impact on the project team's culture than the converse of the project team influencing the organization.

At the same time, the potential impact between the project team culture and other subcultures will be dependent on many factors. Some of the factors include the relative power between the subcultures, amount and intensity of interaction, and the set of potential common problems that they may share. Each subculture may be very unique and the overall impacts and influences are situational dependent.

SUMMARY

This chapter discusses interactions of culture within the context of national and organizational cultures as well as subcultures within the organization. Objectives of this chapter included introducing the reader to the range of cultures which exist, how the various cultures interact and influence each other, and how different power levels have a moderating impact between cultures and subcultures.

An outcome of this chapter is identification of social and organizational relationships which exist within the open system business architectures. Analysis of how the different cultures coexist results in a logical conclusion, which is supported by various literature sources, that national culture can and does influence an organization's culture. This is a direct result of how organizations exist within a nation and how an organization's culture interacts with the nation's culture and is subsequently influenced by it.

The conclusion that the national culture will influence the organization's culture is supported by the fact that the members of the organization are also members of the nation. In this situation, the members' national culture is brought into the organization. Yet, while the organization is influenced by the national culture, it still has and does develop a culture which reflects its own assumptions, attitudes, values, and norms. In this context, it is recognized that the "… dynamic processes of culture creation and management are the essence of leadership and make one realize that leadership and culture are two sides of the same coin" (Schein 2004, 1).

This chapter also expands on earlier descriptions and definitions of organization subculture. As presented, organizational subcultures develop as most businesses consist of distinct groups, teams, or both, which have their own unique business support functions. The personnel in these smaller entities generally have specific and distinctive roles, responsibilities, tools, techniques, educational background, different communication interaction requirements, and different decision requirements, and they encounter different problems. As the members of these groups and teams collaborate and interact, a group or team-specific subculture emerges.

The dynamics within and between subcultures are greater than dynamics at the organizational or national levels. This can be viewed from a system perspective in that each subculture is an open system which receives energy from and sends information back to the organization, as a whole, and to each subculture which it interacts with. The amount of culture shift, modification, or change, which occurs through the various interactions, is moderated by the relative power structure which exists between the interacting entities. That means that the subculture will be influenced, more so, by the organization than the organization by the subculture. This is driven by the fact that the organization has greater power and authority than the subculture group or team.

A similar view occurs when looking at how different subcultures interact and which has the greater impact on the other. Those subculture groups and teams that have the highest power and authority generate greater influences on those subcultures they interact with of lesser stature. Of course, if subcultures within an organization have little to no interaction then little to no subculture influence occurs.

The chapter also expanded the discussion on project teams and their subcultures. The discussion highlighted that project teams, within a common organization, can be homogeneous, heterogeneous, or ad hoc. There are various influences which form or shape the project team's culture and it is essential for the project manager and organization's leadership to understand the shaping forces which are in play.

In the end, research has shown that organizational change and process development is impacted by the various organizational subcultures (Müller, Kraemmergaaard, and Mathiassen 2009, 597). The literature also identifies that an organization's culture is a source for project support as well as a source which can negatively impact project success. Chapter 5 expands on the discussion of project team subcultures and how the team members, the project manager, and the organization's management interact, change, modify, and study it.

DEVELOPING AND MAINTAINING AN EFFECTIVE PROJECT TEAM CULTURE

If, on your team, everyone's input is not encouraged, valued, and welcome, why call it a team?

—Woody Williams (2013)

5.1 INTRODUCTION

Previous chapters have provided the reader with a range of culture discussions such as what are national culture, organizational culture, organizational subcultures, and project team cultures. Each of these was considered and discussed from many different perspectives. The relationship between culture and systems science has also been presented as one method of understanding how different cultures can and cannot influence or impact other cultural environments.

The previous chapters also presented how difficult it is to study culture. This difficulty is based on multiple aspects such as culture is so deeply rooted within our psyche that we are not always aware of why we take certain actions, respond in specific ways, or make decisions based on a specific process. It is the combined influences—culture is often of an unconscious nature, which is deeply ingrained, and how it is often invisible to the individual—that restrict and limit the researcher's ability to directly measure and evaluate culture in a quantitative approach (Henrie 2005, 10). Another factor, which makes the study of culture difficult, is the lack of definitive metrics which the researcher or project manager can use to directly compare different cultural settings or to measure cultural change processes. As an example, we can relate to how hot or cold something is based on a universally acceptable temperature scale. Within cultural research no such universally accepted cultural measure has been developed and agreed to. Measuring culture tends to be qualitative rather than quantitative. To minimize the difficulty in directly measuring culture one must look at the problem from a different direction, that is, the project manager must use indirect indicators (qualitative) rather than direct, quantitative metrics.

Along this line and as previously discussed, to develop an understanding of a team's culture the individual studying the team, be it an engineering manager, some external entity, or a project manager, must focus on identifying indicators which provide a view of what culture is but are not direct measures of culture per se. Some of these indirect indicators include variables such as how the team heroes are established; and what the team's values, rituals, communication styles, artifacts, and shared assumptions are. Other culture indicators that can be analyzed include how the team works together to resolve issues and make decisions as well as how internal and external team communication occurs.

This chapter transitions the culture discussion away from the general culture framework toward a focused dialogue on project team culture. The objective of this focused view is to provide a sequence of steps that one may follow to enhance or to build an effective team culture. The series of steps starts with developing a vision of what the desired project team culture needs to be and ends with the process of monitoring and managing the culture transformation throughout the project life.

While this chapter provides the reader a discussion of some common analysis alternatives, methods, and tools, it is not intended to present an all-inclusive set of approaches that one can use to analyze the team culture. While other analysis approaches are available, those presented here provide a set of models which the project manager can use to evaluate the current team culture state, develop and implement a culture transformation, and take steps or approaches to monitoring and maintaining the project team culture throughout the project life cycle.

Fundamental to this discussion is that while culture is an abstraction, its impacts and influences are essential and key attributes of a successful project team. It is also a fundamental aspect that an effective team culture involves everyone on the team working together to develop the shared experiences which ultimately shape and sustain the team culture.

5.2 WHY IS PROJECT TEAM CULTURE IMPORTANT? ISN'T IT ENOUGH THAT I DO MY JOB, WHY DO I HAVE TO WORRY ABOUT THE TEAM'S CULTURE?

The ultimate objective of all projects is to successfully deliver the intended final results. While defining what project success is, is outside the scope of this book, it suffices to say that every project is unique and what constitutes success is linked to the definition and environment of each project as well. Yet, while sidestepping the definitive definition of project success the following highlights the cultural literature view on what are project success, team leadership, and risk management. As will be noted in more detail, each of these project aspects is linked to culture and from this common culture the team can become highly effective.

From a project success culture implication perspective, the preponderance of literature sources identifies culture as a major contributor to project success. As one literature source discusses, project results, deliverables, and final project outcomes are directly related to culture (Shore 2008, 6). A different literature source presents that the right project team culture frames a positive communication structure that is essential for a project to be successful (Yazici 2011, 21).

The literature has also identified cultural variables such as the leader's values and basic assumptions as critical factors toward the team's culture and success. One case analysis identified that the leader's values can become the catalyst which drives the formation of the team culture. The very way the leader communicates, defines success and failure, as well as how he or she resolves problems become embedded in the team as its nucleus culture (Aronson, Shenhar, and Patanakul 2013, 50).

The literature also identifies that the leader's attitudes, values, norms, and basic assumptions can be a significant influence on developing and sustaining an effective project team culture. An analogy to this comes from the entrepreneur world where the following statements and positions are presented:

1. "... a small firm's orientation is grounded in the values, intentions and actions of the individual who is in charge" (Altinay 2008, 112).
2. "... culture plays a role in economic and social activity" (Minniti 2009, 80).

An essential key is that the leader's culture can impact the team's overall culture. As one source notes, a successful leader's culture can be and often is assimilated within the team which improves the team's environment (Kuhlmann 2010).

Examples of a successful culture assisting a team to succeed can be found on virtually every team sport field. Those teams that share a common set of heroes, problem-solving approaches, and norms can be seen as working teams where everyone knows what to expect of their team mates. Conversely, those teams that lack a common culture appear to be dysfunctional and a group of individuals rather than a team.

The preceding paragraphs presented how project culture is a key component of the project which contributes to project success. It must be emphasized that an effective project culture is required to enhance the probability of project success. Conversely, a dysfunctional project team culture can quickly and easily doom a project. Expanding on the leadership and project culture relationship, the literature discusses how the right project culture is also essential to project risk management or, as it is sometimes called, uncertainty management.

Projects involve risks. This is an explicit characterization of all projects that are grounded in the fact that as the project's outcome is to provide the client some unique product or process, there are areas of uncertainty in all projects which become risks. In this unique environment the project team, at some level, is always proceeding in unchartered waters where risk or uncertainty (here forward uncertainty) is present and where it needs to be effectively managed. "Uncertainty management is recognized as essential to tackle the inevitable uncertainty associated with ... projects ..." (Karlsen 2011, 241).

Yet, while the need to manage project uncertainty is an acknowledged key element, in the underlying project structure, the literature identifies a less than stellar uncertainty management success rate. Specifically, "... recent studies have raised a concern regarding the effectiveness of uncertainty management ... [because] many organizations turn to this activity without understanding its underlying ... culture" (Karlsen 2011, 241). This statement highlights the need for an effective team culture if the team is to be successful in managing its uncertainty and the associated risks.

In summary and to answer this section's lead-in question, the project team culture can either contribute to or hinder the team's ability to be successful. This same culture can also contribute to the team's ability to manage project uncertainty and its associated risks. Conversely, if the project team fails to develop a functioning team culture this will greatly increase the probability of project failure. As project team culture is very fundamental to project success and project uncertainty management, it is essential that the project team members understand what project culture is and how a successful project culture can be designed, implemented, and continuously monitored.

5.3 DEVELOPING A PROJECT TEAM CULTURE: WHERE TO START?

Now that we have built a firm understanding that an effective project team culture significantly contributes to the project's potential for success and uncertainty management, this section begins the dialogue on measuring, implementing, and monitoring culture change. The objective is to answer the basic question: How to develop and sustain an effective project culture? This question arises regardless of whether you are a new project manager, a seasoned professional, if you are just initiating a project, or if the project is well into implementation.

The first step in this process is to determine what an effective project team culture would be for the project's physical and social environment. Then the project manager and project team must take the next step of developing an understanding of the current project team culture. With these two information points understood and documented, a gap analysis can be performed (the third step in the process) which will identify the areas that are working well, areas that may need some tweaking, and areas where major transformational change is required. The outcome of the gap analysis provides input to the development of the transformation plan which ultimately leads to the transformation process. Throughout this effort is the continual process improvement aspect of monitoring, measuring, and adjusting as required.

The culture research literature discusses several ways to approach developing information on the preferred culture attributes as well as an understanding of what the current culture is. Some approaches to developing an understanding of the current team culture include the application of qualitative assessments, quantitative statistical-based surveys, as well as a triangulation method. The following sections provide an introduction on what and how these approaches are used.

5.4 QUALITATIVE RESEARCH

Defining qualitative research is challenging as a host of books, journals, as well as magazine articles freely apply the term qualitative research and provide an equal number of definitions. Some of these definitions are based on an epistemological discussion, others define it by describing how to apply various qualitative research methods, and others provide a comparison and contrast to quantitative research. A simple definition, which highlights a principal qualitative research distinction, is that "Qualitative research involves any research that uses data that do not indicate ordinal values" (Nkwi, Nyamongo, and Ryan 2001, 1).

Thus, the qualitative approach or method develops data which can be assigned various variable attributes. Yet, the assigned variable values have no specific interval assignment. In the end, this approach is used to develop an understanding, but not an exact metric, of the project team's human element, such as culture and human behavior. It is geared to understanding the *soft* systems, such as how the team generally communicates or the values it demonstrates, versus the quantifiable corporate actions such as financial records, lost time accident metrics, or the number of days since the last vehicle accident.

To develop a better understanding of qualitative analysis, it is worthwhile viewing the overall process. Generally, qualitative research starts by observing some event or some outcome and then the analysis proceeds back in time. Moving back in time is intended to develop an understanding of the process which created the observed state and to explain an event based on the cause. The process is referred to as a cause-of-effects analysis; at each stage of the analysis, the review identifies the previous cause which resulted in the observed effect which then leads the analysis to the next initiating event. This train of observation and cause identification is intended to lead the researcher to the ultimate source or answer (Mahoney and Goertz 2006, 230). Qualitative cause-of-effects analysis is very consistent with and frequently applicable to the project environment.

Using qualitative research, the project manager can develop an understanding of what has occurred by tracing the sequence of events back in time until he or she develops an understanding of what caused either the positive or negative behavior. This is the causes-of-effects explanation approach. If the project manager sees a common trend in that when Cause A occurs then Effect B will occur, the project manager and project team have critical and actionable information.

An example of this would be the project manager identifies that numerous engineering design package revisions are required during the installation process. By working backwards the project manager identifies that the observed process is consistent with the majority of engineering packages being received, Moving backwards in time, the project manager identifies that the original drawings, which the design was based on, did not reflect the actual field as-built state, major issue number one.

The next step in the analysis identifies that the engineering design group also failed to verify the state of the drawings before beginning to make design changes. Therefore the project manager, using qualitative methods, has identified that the current as-built drawings are incorrect, no field visits were performed to verify the drawings prior to beginning the design change, and the project team field personnel, while they have time to fix the incorrect design, did not have time to identify the design flaws prior to installing the system.

As this example demonstrates, understanding the actionable processes which occur throughout the sequence of events, allows the team members to either take action which supports and promotes a positive occurrence, until it becomes a basic team assumption, or they can take a different approach, such as not taking any action, to prevent the cause-of-effects stream from occurring. The later action is intended to prevent cultural adoption of an undesirable situation.

A qualitative research caution is the risk of dealing with small populations and small sample sizes. Analyzing one event may determine one causes-of-event result but a slightly different causes-of-event analysis may derive a different cause. What this is intended to identify is that qualitative research, especially in small sample sizes, is fragile where the developed understanding can be quickly invalidated by some slight different event (Mahoney and Goertz 2006, 238). One must be very careful on *reading* more into the causes-of-events sequence than what really occurred.

5.5   QUANTITATIVE RESEARCH

Quantitative analysis is often described as the converse of qualitative research. This comparison occurs as qualitative research follows an effects-of-causes approach rather than a causes-of-effect analysis. Quantitative analysis also relies on statistical techniques which require the use of interval variables. As previously noted, one cannot derive any meaningful statistical information from categorical variables; therefore, quantitative analysis requires the ability to measure events using interval data.

Interval data involve two or more values which are evenly spaced, ordered, and have specific levels of measurement. Obtaining the weight of potato chip bags, which are being filled on the assembly line, is an example of obtaining interval measurements of the bag's weight. Based on the interval values obtained, various statistical methods can be applied. The statistical results allow the development of meaningful information about how the potato chip bags are being filled, sources of errors such as under- or overfilled bags, and the rate of bag fill.

Quantitative analysis is also often characterized as controlled experiments. That is, an attempt is made to keep everything identical throughout the event except the variable of interest. In the potato chip bag example, the variable of interest is: How much does the bag weigh after filling? While gathering the intended data other potential variables of interest, such as the size of the bag and the type of potato chip, are kept constant. Using this approach allows the researcher to develop meaningful statistical bag weight results such as the average weight and the standard deviation for the measured group of bags. The information provides quantifiable information which can then be used to monitor the system quality metric of filling the bag to a certain weight plus or minus a specific standard deviation.

Within the project environment, the application of quantitative analysis which is intended to derive an understanding of the project team's culture is challenging to accomplish. There are many issues to this approach such as obtaining a sufficient data sample size to derive any meaningful statistic. That is, many project teams are small which automatically limits the number of quantitative analysis responses. Further, and as has been noted several times, most projects are of short duration. Further, there are limited means of developing quantitative metrics when dealing with culture. Coupling the small sample population size, short project duration, and few directly measurable culture interval measurable variables restricts, but does not eliminate, the use of quantitative research methods.

An example of a project team quantitative analysis which has been and continues to be applied is the use of a survey to determine the team member's power distance index (Earley and Erez 1997). This tool is intended to identify the participants' power distance index which identifies how the team members respond to direct authority (177). Understanding how the various team members respond to authority is a key culture attribute. If there is disparity in how the team members respond to authority it can be a major source of project team issues. Using this survey tool, or similar survey tools, provides a means to determine a cultural indicator.

5.6 TRIANGULATION

A challenge in identifying a project team's culture is the need to infer the results based on indirect variables. The previous sections discussed how the project manager could use qualitative or quantitative approaches to determine or infer the team's culture based on processes such as direct observations, questionnaires, and surveys. The problem with these methods, within a project team environment, is that the project manager generally deals with a small number of participants, few examples, and short timelines. The project manager is also part of the environment and subsequent culture. The outcome of this type of analysis may be individual data points which may or may not provide a clear indication of the inferred team's culture.

The problem of small sample sizes and data sets is not unique to the project management discipline. It is an issue in many other analysis areas. To help address this issue, a different analysis approach has been developed named triangulation.

Triangulation is similar to the old seafarers' method of determining their location on the sea using celestial navigation. In the seafarers' case, they could determine their approximate location by taking site readings on multiple stars. Each star provided a level of information that by itself did not provide a clear indication of where the ship may be. To better define the ship's location, other star site readings are obtained and when combined together the ship's location is identified. A similar approach applies to analyzing the project team's culture indicators as well. The project team's triangulation analysis starts by assuming that obtaining several observations of the same culture environment allows one to develop a better understanding of the culture under analysis than a single observation allows (Stake 1995, 110). That is, as with the sea analogy, each observed or derived data point adds more information which helps to develop supporting information for the answer or disprove or in some way modify the earlier assumptions and concepts.

As an example, the project manager determines a cause perpetuated the observed effect based on one event. This one event could lead the project manager to determine how the project team was interacting and indirectly inferring what the cultural underpinnings are. Yet this is a single point on the wide body of the project team's potential cultural ocean. To provide further validation or invalidation of the first analysis, the project manager would determine the cause of a similar but different event. If this second point closely matches the first, there is support for the initial view. Then, if possible, the project manager could conduct a small quantitative data sample analysis to see if the outcome of this process provides further support or drives him or her to alter or modify the earlier conclusion. Each incremental data input adds information which guides and supports development of the answer.

The essential triangulation key is to derive a view of the project team's culture from different perspectives using different methods and approaches. Each new perspective provides further information which aids in clarifying the inferred culture team foundation.

5.7 ETHICS

In the process of analyzing, planning, and implementing project team culture changes, the project manager must be fully cognitive of the ethics which surround these processes. Ethical behavior is essential and vital to any organization at any time but especially when they are trying to change how the team interacts, communicates, and makes decisions.

Popular literature, research articles, and news reports provide ample corporate examples where the loss of ethical behavior resulted in dramatic negative company impacts. There is the infamous Enron ethical scandal that destroyed that company. Then there is the Barclay's ethical scandal where they admitted to artificially changing London Interbank Offered Rates. While this admission did not cause Barclay to implode, it did result in hundreds of millions of dollars of fines. There are also cases where project managers failed to implement or follow ethical behavior which resulted in tremendous negative impacts on the project team as well as cascading impacts within the organization.

Putting ethics in context, the study and discussion of ethics originated with ancient philosophers such as Aristotle, Confucius, Heraclitus, Plato, and Socrates. From these early efforts, throughout history, and into modern times, the study and application of ethics has been a continuous process and one with different connotations. One definition of ethics is "the principles of conduct governing an individual or a group" (*Merriam-Webster* 2013).

Ethics are also described as an individual as well as a group behavior construct on what is viewed as right or wrong (Fellows, Liu, and Storey 2004, 289). Other definitions of ethics link the terms ethics and morals together where morals are described as those things which guide interactions through a set of rules and principles (288).

Ethics are cultural based and exist with distinct environments. This means that what is ethically acceptable within one society may not be ethically accepted within another. An example of this is the application of nepotism. In some nation-states, nepotism is not only culturally accepted and viewed as a positive ethical practice but it is virtually required. Other nation-states have a cultural and ethical view that nepotism is wrong and the members of this society take a very dim view of it when it occurs.

Yet, how does this apply to the project management actions around the project team culture? The overarching answer to this involves the purposeful intent to alter the team, and subsequently the team members' culture. It also is directly related to the general project management context which involves taking actions or imposing sanctions on the group. These actions must occur within an ethical and morally accepted framework as they directly affect those within the team as well as the broader organization (Helgadóttir 2008, 743).

Acknowledging that different societies may have different ethical underpinnings only strengthens the position that the project manager must be aware of these factors and act accordingly. It is essential that as the project manager and the associated team analyze the team's culture as well as develop methods to alter or sustain it, they must be very aware of the ethical component of what they are doing.

As the project manager and the management team enter into and implement a culture change control process, there must a firm understanding that management:

1. Is responsible "... to make sure processes are just, fair, and reasonable and do not violate human rights
2. [Is] responsible to maximize the overall utility for the stakeholder"
3. Should exhibit exemplary conduct stemming from stable dispositions to act." (Müller et al. 2013, 30)

In summary, ethical behavior is an essential cultural trait and one which the project manager as well as the project team must foster and support. To assist in this effort, several professional organizations have developed and do provide ethical codes of conduct for their members to follow. The project manager and project team must avail themselves of these resources for assistance in this very critical area.

5.8 TOOLS

The preceding paragraphs provided a view of different analysis methods which generally apply to developing an understanding of the project team's current culture. This section contributes to the general method discussion by discussing evaluation and analysis tools which the project manager may elect to use rather than develop his or her own approaches. As such, what tools are available to assist the project manager in benchmarking or understanding his or her project team's culture? One answer to this question is the leveraging of available maturity models.

A maturity model is also commonly referred to as a benchmark or service mark model. Regardless of terminology, the intent is to provide an approach or model which allows the project manager to develop an understanding of the organization's maturity for a given process. Maturity models fill several basic functions. First, it provides a means to evaluate an organization's capabilities using a common format and approach. The maturity model's standard format and approach bring structure and repeatability to the analysis effort. Second, it provides a means to compare an organization's capabilities against their peers and competitors. As the model is a structured instrument, it provides the evaluator a means to compare different project team evaluation results within a format which provides comparable information.

Third, it provides a means of identifying the characteristics and attributes of the organization's maturity level. This information identifies the areas that may be working well and points to processes where improvement maybe beneficial. Fourth, it supports the organization in developing change processes that are intended to advance the system to the next maturity level. Fifth, it provides a consistent mechanism to monitor the organization's capability on a continuous basis.

One way to visualize the maturity model is shown in Exhibit 5.1. As identified, for the organization to transition from a neophyte or novice to an optimal level is similar to climbing a ladder. Each step or process leads you higher up the maturity model and with every step up you rise higher and perform better.

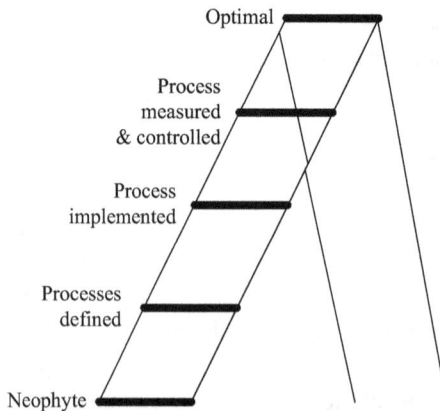

Exhibit 5.1. An example of a maturity model.

The intent of Exhibit 5.1 is to identify that a capability maturity process is a continuum which spans from virtually no processes or understanding in place to ultimate or optimized capabilities. Each step along the way brings more value to the project and to the organization as a whole.

Approximately 30 different maturity models have been developed since Carnegie Mellon 1986 to 1993 Capability Maturity Model formation (Brookes and Clark 2009). A few of the available project discipline specific maturity models include:

1. Organisational Project Management Maturity Model (OPM3®), the Project Management Institute
2. Project Management Maturity Model (PMMM^{SM}), the PM Solutions
3. Cabinet Office's Portfolio, Programme and Project Management Maturity Model (P3M3), The Cabinet Office of Rosebery Court, St Andrews Business Park, Norwich, NR7 OHS
4. PRINCE2® Maturity Model, The Cabinet Office of Rosebery Court, St Andrews Business Park, Norwich, NR7 OHS

As noted, each of these, as well as other available maturity models, focuses on the project discipline; the question becomes, how can one use a maturity model to measure the project team's culture?

The answer to this is grounded in the concept that measuring a team's culture involves measuring indirect indicators. As culture is not directly measurable, the analysis must rely on variables which can be measured while ultimately providing information about what the underlying culture is. Some of the available maturity models provide information on these indirect indicators. One example of this is identified in the Project Risk Maturity Model (RMM) literature which states:

> The Project RMM contains 50 questions, each one of which can yield information about a project's risk management process ...
> Since risk owners are responsible for managing their risks, the answer[s] ... will yield information about whether or not risks are properly understood ... and whether or not the project has a good risk management culture .... (Hopkinson 2010, 7)

As noted, Project RMM not only provides information on the project team's understanding of risks but an outcome is identification of the team's risk management culture as well.

As identified and discussed, utilization of maturity models is one means of delineating the current project team's culture as well as providing a benchmark as to the current state. This information is critical input to the next phase of planning, implementing, and monitoring the culture transformation continuous improvement process.

In addition to the various project environment maturity models, there are other tools which the project manager can leverage in his or her efforts to identify, monitor, and plan cultural transformations. One such tool is referred to as an environmental assessment tool. There are several examples of environmental assessment tools which are available to the project manager. As an example, one is provided by the World Bank. This tool and assessment method is "... used in the Bank to examine the potential environmental risks and benefits ... It is an essential tool for integrating environmental and social concerns ..." (World Bank 2013). This tool provides a means of assessing the project's social concerns and culture environment.

Another example is the Project Environment Assessment Tool (PEAT©). Originally developed in 1997 it is designed "... for measuring and determining the environment that supports projects success ... [and] to determine how well organizations support project management" (Hoole and Plessis 2002, 256). PEAT evaluates the factors:

1. Upper management support
2. Project planning support
3. Customer or end user input
4. Project team development
5. Project execution support
6. Communication and information systems

This tool, as other tools, is designed to provide end users a means to assess their organization's support for project management, identification of gaps which may exist, and provide a benchmark to other organizations. As noted in the literature, this tool can be used "... as an evaluation tool for project success or as a process to create project culture in the organization" (Hoole and Plessis 2002, 264). The key point of these tools is the process of examining those indirect culture indicators as a means of understanding the project team's cultural attributes.

The preceding text is not an inclusive list of tools or methods which are available to the project manager or organization leaders. Many other tools are available to the interested parties as well as commercial service organizations that have complete culture transformation processes to offer. These commercial service providers' programs, methods, and tools are intended to assist project managers in understanding their team's current culture, benchmark it against their peers, and assist in developing transformation processes. Project managers should evaluate the various available methods, tools, and approaches and select an appropriate path forward which supports the objective of developing a holistic culture transformation process that is appropriate within their organization's structure and their unique project environment.

5.9 DEVELOPING A CULTURE TRANSFORMATION PLAN

Developing a culture transformation plan may not be required in your specific project. You may find that the project members quickly and effortlessly become a team, and naturally develop a holistic culture which supports the overarching project success objective. But, what if you are not that project manager and you determine that your team must alter the current culture to enhance the probability of project success? In this situation, a transformation plan must be developed, implemented, and monitored. This section provides one approach to developing such a transformation plan.

To start with and before any change can occur, project managers must have a clear vision of what their unique project culture successful attributes are, that is, the desired end results. They must also understand what the current, or as-is, culture state is. With an understanding of the desired culture end state, and the as-is culture state they can then identify the gaps between these states. With this understanding, the details of the transformation plan can be formed. Throughout this process, it is also paramount that project managers recognize and accept that they, as well as "… other managers (leaders) of the parent organization … [are] responsible for arranging conditions conducive to a creative and disciplined culture …" (Cleland and Ireland 2002, 571). This means that they must take ownership and responsibility for the overall process, and structure the transformation within their specific physical, social, and culturally unique project environment.

As such, it is clear that you must start with a vision of what constitutes critical attributes of your unique project team's successful culture. But what are these attributes, how do you identify them, where do you start, and what is required in the vision?

As with the study of culture in general, a successful project team culture is unique to the project's physical, organizational, team structure, and social environment. It includes what occurs internally to the team as well as external influences, such as the organization and national cultures, which are part of this open system. One project team's successful culture attribute set may not be the same set for a different project team environment.

As an example, if a project is implemented in an environment where the organization and national culture supports, encourages, and even requires individual project team members to perform as equals, that is a low power distance index project environment; the successful project communication culture is different if the converse project environment were encountered. In the low power index environment, the successful project communication culture attributes include aspects of open and frank, bidirectional communication exchanges. Isolation and restriction of forthright dialogue would not be part of the accepted culture.

Conversely, if a similar project were to occur in a high power distance environment, where a rigid and strict subordinate to superior relationship is the norm, a different successful project culture is needed. Within this environment, it is unrealistic to develop a project team communication culture attribute vision which includes open and forthright bidirectional communication exchanges as that culture is diametrically opposed to current reality at the project, organization, and national levels. This may be driven by the environment where "Relationships between subordinates and superiors ... are frequently loaded with emotions" (Hofstede 1997, 36). This emotionally charged environment may drive the project team culture communication vision to focus on how peers communicate with each other and how the hierarchical communication will flow rather than one of openness, frankness, and equality.

The preceding case is just one example which demonstrates the need to be cognizant of the project environment when developing a successful project team culture vision.

One approach to developing this vision is to work from a culture checklist which assists the vision developers in reviewing and documenting pertinent components of this vision. One form of the checklist could be based on the following outline:

1. Symbols or artifacts
   a. What are the project symbols which would be identified as indicators of this successful project?
   b. Should the project have a unique logo and phrase which clearly identify the project team as a unique team?
   c. Should the project team be collocated in a central area?
   d. For a collocated team, how should the physical environment be structured? Should everyone be in a cubicle with many open meeting areas or would the use of offices be more appropriate?

e. What about providing the team members a project shirt, pen, backpack, or some other physical item which clearly says they are a member of the team?

2. Rituals
   a. What rituals does the team see as being part of the successful project team?
   b. Should there be a ritual when people join or leave the team?
   c. Should there be a ritual when a successful major milestone is met?
   d. Should the project team have a meeting ritual such as all meetings will start with a safety moment or some other critical vision of the project?

3. Heroes
   a. What constitutes or would constitute a hero for this project?
   b. Is the hero defined as someone who puts in the extra mile?
   c. Is the hero defined as someone who fosters collaboration and inclusion?
   d. Is the hero defined as someone who exhibits the desired values?

4. Values
   a. What specific mode of conduct will be acceptable on this project?
   b. Will the team value system include the ability to have open and frank discussions?
   c. Will the project team and the management structure accept bad news without *shooting* the messenger?
   d. Are the team's values based on what is best for the team or the individual?
   e. What value system will be applied to conflict management?

This list is obviously not fully inclusive but provides a foundation from which the project manager and their team can begin. By explicitly and systematically working through the list of culture attributes, the team can define and document a culture vision.

Having a vision of the project team's culture needs is essential before a plan can be developed. An analogy is all roads leaving a city go somewhere but only one road will lead you to where you really want to go in the shortest time. You must know what the destination is to plan the travel to the destination. But, knowing where you ultimately want to be is only part of the picture. Knowing and understanding the project team's as-is culture is the second piece of this puzzle.

Determining and understanding the project team's current culture is easy to say but can be difficult to determine. Several ideas and methods have been proposed and applied on how to do this across many different environments and cultural contexts. These efforts have met with varying levels of success and failure. To provide an increased probability of being successful requires the project manager to understand precisely what he or she wants to achieve which includes the ability to present it to both senior management as well as the project team. This is the first major challenge for the project manager to overcome and is very project specific.

Once what is to be achieved is understood and stated, the discovery of the current culture state can occur. One approach to this would be the utilization of Schein's 10-step intervention. This intervention consists of:

1. Obtaining leadership commitment
2. Selecting groups to interview; in the case of the project this would be the project team
3. Selecting an appropriate setting for the group interview
4. Explaining the purpose of the group meeting
5. Presenting a short lecture on how to think about culture
6. Eliciting descriptions of the artifacts
7. Identifying espoused values
8. Identifying shared tacit assumptions
9. Identifying cultural aids and hindrances
10. Reporting assumptions and joint analysis (Schein 2004, 340–7)

An aspect of this process requires that someone other than the project manager lead and conduct the assessment. This is based on the assumption that by not having the project manager present, it provides for more open dialogue and information exchange between and within the project team members than if the project manager is present. In most situations, the presence of management tends to limit how frank and open the team members are willing to be. This is a cultural artifact which, ideally, should be changed.

A slightly different approach involves the process of developing a

"... list of what comprises a culture: traits and behaviors, assumptions and beliefs, values and norms, language utilized (not national language but organizational language), rituals and customs, socialization and sub socialization, how problems are solved, tools and technology used, and finally the layout of work areas." (Lane 2013, 52)

The key is to develop an understanding of the as-is culture by *fleshing out* these observed project team environment attributes.

Another approach is to develop or leverage an existing culture survey which is supplied to the project team. There are several culture-based surveys which are available as either stand-alone efforts or part of for-profit consulting firms' methods. In this process, the project team members would anonymously complete the survey; at which time the project manager or contracted consultant would analyze the results, consolidate the information, and then meet with the project team to present the outcome. The outcome of this effort becomes the foundation for developing the transformation plan.

In essence, there are several different approaches to identifying the as-is project team's culture. The project manager and their executive team should identify a method and tool or a set of tools which provide input within their unique environment. Throughout the process, a deep awareness of the project team's attitudes, feelings, and comfort level is required.

With a clear understanding of the ultimate objective and the as-is state, the next step is to identify and detail the "... desired traits and behaviors vital in supporting the strategy; create awareness of leadership's current traits and behaviors; and identify opportunities and actions to align difference between the desired and current behaviors" (Lane 2013, 49). This is the gap analysis and development of the gap elimination plan process.

As each project environment is unique, the detailed gap analysis and subsequent gap elimination plan must also be unique to that project. In general, the process is to first develop an understanding of the difference between where we are and where we want to be. This understanding must go beyond general statements to one of specific details which include actionable items. An example may be that the current decision-making process is primarily ad hoc with each person doing the best he or she can as a stand-alone entity. The desired culture decision-making attribute is that decisions will be made with the input and contribution of all involved team members. Further, the desired decision-making process is one where all participants can challenge the proposed decision, make suggested changes, and assist in the active development of a cohesive answer. The actionable outcome is changing the project team members' decision-making process that all future decisions will be an inclusive, not exclusive, process.

In this example, the gap is the difference between who is included in the decision-making process as well as how the decision is made. To close this gap, the culture change management plan would detail who is involved in which types of project decisions as well as the collective nature of making the decision.

In essence, development of an actionable culture change plan involves the process of (a) defining explicitly where you want to be, (b) defining, detailing, and understanding the as-is culture attributes and variables, (c) unambiguously detailing the gaps between the here and now and where we want to be, and (d) developing a detailed plan on how to close the gaps as well as how the process will be managed and monitored. The detailed plan must include specific, actionable, tasks and activities. The danger is that the plan developers revert or leverage high-level, generic statements, which look and sound great but fail to provide any firm action requirements.

Developing the overall process is not simple or easy, but the literature is clear that it provides significant benefits if properly implemented.

5.10 TRANSFORMATION PLAN IMPLEMENTATION AND MONITORING

The preceding information is not the project team culture change process silver bullet which can be applied unilaterally and universally to all projects, in all environments, to develop a project team which is always successful. Yet, the information provides the project manager a deeper understanding of the reasons why project team culture is important, some tools which may be useful, and outlines a general approach to understanding the as-is project team's culture as well as how to plan the team's culture transformation change process. It also stresses that need to develop specific actionable tasks and activities which provide support in achieving the desired end result.

As was presented earlier, to increase the probability of success it is essential that the project manager and their management structure have the vision and commitment to implement and manage any culture change process. It is also imperative that they understand that implementing and managing culture change are not easy or simple. Often the project team is facing the need to change habits, tweak the working environment, change how they communicate, and alter how they engage in communications, risk planning, and decision making. Change is scary and many people resist it as long as their level of the as-is state of discomfort is less than any discomfort that will occur during the change process.

To overcome the resistance to change requires establishing conditions which involve a combination of processes, events, and vision that are conducive to a creative and disciplined change process. One approach in creating a conducive environment is the utilization of rhetorical dialogue. Rhetorical dialogue is the use of oral and written information which is intended to inform, persuade, and motivate the project team members toward a common culture. Rhetorical approaches include but are not limited to the use of:

1. Analogy—A figure of speech which presents the reasoning or explanation between two different events, things, or ideas which highlights the similarity of either the desirable or undesirable trait. One example of an analogy would be where the project manager likens the process of monitoring the risk management plan to managing your investment portfolio. In each case, the intent is to identify events prior to their occurrence such that proactive efforts can be taken to either mitigate the negative or enhance the potential outcomes.

2. Chiasmus speech—A figure of speech in which the second half of an expression is balanced against the first with the parts reversed. As an example, when the going gets tough, the tough get going, is a chiasmus. The use of chiasmus provides the project manager a means to leave the project team members a strong impression and something to remember as a means to foster change.

3. Irony—A figure of speech which says one thing but means another. It can be used to enhance an understanding by expressing the opposite. One example of irony would be when the project manager claims that as the project team is failing in its internal and external communications it is assuring that the project will be successful. The key is to use rhetorical irony communications to stress the importance of and need for acceptable culture-based communications by actually saying the opposite.

4. Maxim—This is an "… easily remembered expression of a basic principle, general truth, or rule of conduct" ("What Are Dueling Maxims?" 2013). A classic example is "Actions speak louder than words." The project manager may use this maxim to stress the importance of action and the need to "walk the talk."

5. Metaphor—This is the active comparison of two unlike things which share an important role. A project manager could use the example that unfettered spending is like not turning the water cooler faucet off. In each situation the limited resource, money in the first and the water contained in the water bottle in the second, is consumed without obtaining the intended benefit.

Along with rhetorical approaches the project manager can augment and enhance the culture change environment through the use of:

1. The carrot and the stick. This process involves rewarding the positive and desirable cultural attribute while negating the converse.
2. Leadership which pays "... attention to, control and reward; their role modeling and coaching; how they allocate resources; how they select, promote, and 'deselect' people; and the organizational structures and processes they create" (Schein 2004, 291).
3. Generation of disequilibrium within the project team. In essence, the project team is placed in a state where the current situation is uncomfortable and undesirable. This state creates disequilibrium among the team members. The project team's disequilibrium forces "... a coping process that goes beyond just reinforcing the assumptions that are already in place" (Schein 2004, 321). It results in a new culture based on learning how to cope with the new environment which supports a new level of project team comfort and equilibrium.

While the use of multiple processes and events provides a means to implement a culture change, it all starts with a vision. The vision not only provides a clear direction and ample motivation but also the underpinnings of a supportive environment (Heath and Heath 2010, 255).

To help overcome the fear of change management, you must first start with instilling the belief that change is required. To achieve this, the project manager must be able to effectively communicate why change is required, that change is not impossible to achieve, and to obtain the project team's acceptance and *buy-in*. This is analogous to creating a new habit where there are "... only two things to think about: (1) The habit needs to advance the mission... (2) The habit needs to be relatively easy to embrace" (Heath and Heath 2010, 216). While ultimately changing and sustaining a project team's culture is more than just creating a new habit, it can start with this simple premise.

It is also imperative to keep in mind that "People are the driving force in successfully accomplishing change" (Suran 2003, 31). As such, people or in this case the project team cannot assist in implementing change if they are not (a) aware of, (b) involved in, and (c) committed to making it successful. Building the foundation for the team's involvement and active participation involves communication. In this light, it is imperative that an effective rollout communication plan must be developed, implemented, and followed up. The communication plan must articulate the quantifiable end result and vision, the gap between where the team is and where it needs to be, and how the change process will occur.

This communication process has multiple roles:

1. Convincing the project team members of the need for change. The project manager must make it personal so that each team member sees clearly the need to change, how he or she fits within the change process, and what the ultimate benefits will be.
2. Obtaining project team member feedback and input. Before the project team members can buy into the process, they must be involved in the process. This includes being able to provide feedback, suggestions, ideas, and even challenges to the plan.
3. Clearly articulating that the change process is endorsed and supported by upper management. The project team must clearly identify that the organization's management agrees with, supports, and is fully committed to this plan.
4. Providing a roadmap of how the change will occur. It is essential that the project team understands the process without being overwhelmed by a perceived insurmountable mountain. Outline and detail an initial set of four to seven activities where early wins can occur. Provide clear information on how the processes will be implemented, what changes are expected, and how the implementation will be monitored. Do not overwhelm team members with infinite details or an extensive and complex plan. Focus on the near term and low-hanging fruit to show positive results early and often.

Once the initial communication has occurred, an important task is to provide the team an opportunity, location, and structure where they can informally meet to discuss and assimilate the information. This is akin to social movements' *free spaces*. "Free spaces often play a critical role in facilitating social change" (Heath and Heath 2010, 246) as it allows the team members the ability to discuss, develop a collective plan, and a means to establish a new language for the new culture. You can only achieve project team culture change if the team buys into it, supports it, and accepts it.

As the project team culture change proceeds, it must be actively monitored to see if the desired and intended results are occurring. This overall program is akin to a common continuous improvement circle which includes plan, initiate, and monitor, as shown in Exhibit 5.2.

One cannot assume that a culture change process can successfully occur by just developing a plan and implementing it. As with any change

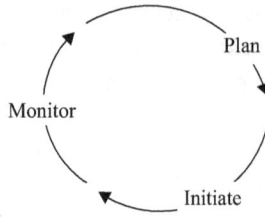

Exhibit 5.2. Culture transformation.

process, the very act of implanting a change impacts the project team which can and will probably impact the plan or how the plan is being implemented. Stated in another way, the plan impinges on the team and the team's response impinges on the plan and implementation strategy. This bidirectional interaction must be monitored and the team must also be willing to adjust the plan as well as adjust the implementation strategy based on the system dynamics.

SUMMARY

This chapter focuses the culture discussion within the project team subculture. The objective of this focused discussion is to lead the reader through a sequence of cultural steps which start with identifying the desired successful team's culture attributes and end with the process of implementing and monitoring the changes.

One key chapter takeaway is the need to understand and quantify what the successful project team culture is. One way to consider this is that all roads out of a city lead somewhere but to arrive at a specific destination requires travel on a specific road toward a specific destination. You cannot implement a project team culture change if you do not know what that desired culture should look like or how to travel the desired path.

The chapter also outlines the fact that once you understand the destination, you need to know where you are today, the as-is state. This involves determining the current project team culture. The chapter provides some general research methods and tools which could be utilized in this effort. The methods and tools supplied are not fully inclusive but are provided as a sample of what is available.

The next phase of the transformation effort is to understand and document the gaps between the as-is and the where we want to be. This gap analysis provides the foundation for developing the transformation plan. As identified, the transformation plan must be attuned to the physical and social project environment. What would constitute a project team's culture-based symbols, rituals, heroes, and values will be dissimilar in different social and project team environments. It is imperative that the project team understand how national and local cultural attributes will impinge on the project team's subculture. This understanding will guide the development of the plan and its implementation which will generate the highest probability for success.

The chapter also identifies that project management is an essential methodology and set of methods for companies to be successful in the world with ever shorter delivery time frames and higher quality requirements. Yet, there are almost endless examples of failed projects and almost unbelievable sums of money spent with negative returns. While there are many reasons for the vast range of project failures, an area which comes as a key focus and directly applicable to the team is management's cultural acceptance and support. Not only must the management structure accept and endorse the project management methodology and methods but they must also support the unique culture which occurs within the project team subculture. "Organizations can successfully manage projects within the traditional functional organization, if the culture encourages cross-functional integration" (Hoole and Plessis 2002, 255). The team cannot operate in isolation if it is to be successful.

CHAPTER 6

CROSS-CULTURAL PROJECT TEAMS AND CULTURAL INFLUENCES

The way a team plays as a whole determines its success. You may have the greatest bunch of individual stars in the world, but if they don't play together, the club won't be worth a dime.

—Babe Ruth (2013)

6.1 INTRODUCTION

This chapter moves the discussion beyond the homogeneous team culture environment, that is a predominately a single culture team, to that of culture within teams which include people from different cultures. The different cultures generally are national based but could be organizational as well. Throughout this chapter, we will refer to this context as a cross-cultural team environment.

The intent of this chapter is to provide the reader some background information and a cross-cultural knowledge foundation of the various aspects which cross-culture teams may exhibit. These various aspects, functions, and interaction tend to be different from teams with a single culture base.

The chapter begins by discussing the difference between multinational and cross-cultural characteristics. As is noted later, the literature tends to use multinational and cross-cultural terms interchangeably. The following discussion is intended to provide the reader insight into these terms.

The chapter then moves into a discussion of the various challenges that organizational managers, engineering managers, project management practitioners, students, and researchers may face within this context. There are unique challenges to developing an understanding as well as working within this cultural environment.

Based on this foundational information, the chapter presents the impacts various cultural attributes, such as communication; power distance; uncertainty avoidance; and individualism versus collectivism, may have within cross-culture teams.

This chapter's objective is to increase the reader's understanding of how interactions of different cultures can impact the team environment and the team members. At the same time, the chapter provides some general guidelines on how managers and team members can work within this context to improve the possibility for team success and individual team members opportunity to experience satisfaction. The team discussion is presented in the context of a project team but is applicable to all teams.

6.2   WHAT IS A MULTINATIONAL PROJECT TEAM AND HOW IS IT RELATED TO CROSS-CULTURE?

To begin forming an answer to what is a multinational project team, let us first review Chapter 4 discussion on what is a country, nation, state, and nation-state. As previously noted, the terms *country, nation, state,* and *nation-state* are encountered throughout the literature and often used interchangeably. The approach by some authors is that these terms have the same definition and thus are interchangeable. Yet, the terms have unique definitions and understanding the difference is very germane to avoiding some confusion when discussing cross-culture project teams.

To reemphasize the distinctions that these terms have, the term country is associated with and refers to a distinct physical location. It is defined by physical borders (Admin 2011). As such, when looking at the world today one sees it divided into approximately 195 different countries. Each of these countries has a distinct geographic border which defines its boundaries and area contained within it. A country, as is expanded on later, also encompasses at least one state but can include multiple nations. That is, a country may geographically encompass multiple nations or even share nations with other countries. Within this context, a country could consist of a single culture but by definition a country is a place on earth which is geographically defined, not culturally established or defined.

A nation is different from a country as it identifies people with a common ancestry which in past times was based on genealogical lineages. Over time and through common application the current meaning of a nation has expanded beyond that of a common blood lineage to be inclusive of those who share a common postulated interrelationship—a *blood* bond between members. In the modern world, the people of one nation may have ancestral or postulated interrelationships. The essential key to a nation is that the linked people hold a strong interrelationship (Rasmussen 2001). People within a nation are said to share a common cultural heritage with all aspects of a shared culture, such as its artifacts, values, beliefs, and norms. A nation is not bounded or restricted by a country's geographic boundaries or a state's structure and authority.

A state is different than a nation but is closely aligned with a country. A state does not require a common culture or language or a strong member association. Rather a state is a system of institutions, structure, and authority which exercises control over the population. It is a social structure which involves entities with legal and legitimate authority. Through this structure a state passes and enforces laws, levies taxes, and establishes forces for the protection of the general population. A state is also different from a nation as its people may not share a common culture.

A state is often directly associated or linked to a specific country. That is, for each country there is a set of institutions, structures, and authorities designed to provide control, protection, and services and functions, for the population, which exceeds the individual or small groups' capabilities. Thus, within a country, there may exist a single or multiple states. Neither countries nor states need to have and often fail to have a single culture base.

The term nation-state is of more recent origin which brings together the definition of nation and state. As such, a nation-state is an independent state inhabited by all the people of one nation and one nation only. Within this context, those who live within a nation-state will share a common cultural heritage, language, and hold a deeply ingrained association. The nation-state will also have a set of institutions, structure, and authority of control over the population. The recent acknowledgment of a nation-state is associated with the migration away from the literal bloodline nation cultural reference to that of a shared culture based on postulated interrelationship, which occurs due to close interaction and development of a shared culture. An indigenous native reservation can be viewed as a nation-state. Those within the reservation share a common culture as well as institutions, structures, and authority of control over those within the reservation boundaries.

With this foundational discussion about countries, nations, states, and nation-states, the chapter advances with the process of deriving a definition of what is a multinational team. To start developing a definition of what a multinational project team is, we look at several defining attributes.

The first multinational team cultural attribute under consideration is the combined elements of size and objective. From a size view, the team must consist of two or more people. This is in full alignment with the definition of a team as well as an underlying feature of how cultures are formed, modified, and changed. That is, culture involves multiple people directly interacting.

This team will also share a common objective of working together toward the achievement of a mutual goal(s) or objective(s). Stated slightly differently, a team is a small set of people who not only share a mutual end objective but also common factors such as an agreed to work approach, performance objectives, and reporting structure. The team also holds each member accountable for participating and performing at a required level of performance (Katzenbach and Smith 1993, xvii). For this book, the common team purpose, process, and performance goals focus on the ability to successfully deliver the project outputs, which ultimately delivers the goals or objectives. Further, within a project team environment the commonly agreed upon working approach includes the project charter, project plan, and supporting documentation artifacts.

Another aspect of a multinational project team definition attribute is that the team members will have complementary skill sets. The actual component of skill sets will be project specific and driven by the scope of work definition. It is through these complementary skill sets that the project team is able to accomplish the variety of activities and tasks required to deliver the final product or service.

Another multinational project team definition attribute is that the team members will include people from two or more nations. This is intuitively obvious as people from different nations are required for the project team to become multinational, which is the multi portion of multinational. Unfortunately, there is a popular or common view that a multinational project team must include people from different countries. Yet, as presented in the earlier discussion, this is not a requirement. A project team could include people from very different nations, which have very unique cultures, while residing within a common country or state. This book's position is that a multinational team attribute is not specifically related to country or state. This position is in alignment with *Merriam-Webster*'s definition that multinational means "… of, relating to, or involving more than two nations" (*Merriam-Webster* 2014) and "Researchers need to zero in on an oft-neglected feature of cross-cultural management—that cultural diversity can exist intranationally or within a single country, as well as across nations" (Jacob 2005, 515) or across countries.

Therefore, this book's derived multinational project team definition is:

> A multinational project team is defined as two or more individuals who are working toward a common goal and who share complementary skill sets. The team will also include people from at least two nations which may or may not include people from different countries or states.

This definition clearly links a multinational project team to one that includes more than one culture. With at least two different cultures, the project team exists within a cross-cultural environment.

6.3 CULTURE AND MULTINATIONAL CULTURE: THE VARIATIONS

Throughout this book, the reader was exposed to various researchers' culture models. These models include Hofstede's five cultural dimensions theory, Fons Trompenaars and Charles Hampden-Turner's seven dimensions model of national culture differences, Edgar H. Schein's organizational culture model which is based on the attributes of artifacts, assumptions, and espoused values, as well as Terrence E. Deal and Allan A. Kennedy's collaborative effort which defines organizational culture within the context of:

- Work-hard, play-hard
- Tough-guy macho
- Process
- Bet-the-company

Specific to the national culture models, literary challenges to their applicability within many contexts as well as to their validity have and continue to occur. As an example, one common challenge, which is supported by various culture model researchers such as Hofstede, is that the national culture models develop results which cannot be extrapolated at the organization or individual level. As one source states, "Hofstede's dimensions of national culture were constructed at the national level. They were underpinned by variables that correlated across nations, not across individuals or organizations" (Brewer and Venaik 2012, 673). One cannot point to any single individual and demonstrate that the overarching national culture is fully replicated or represented in that person.

Another way of viewing this challenge is along the statistical lines of an average. When one takes a data set's average, 50 percent of the data is below and 50 percent is above the average value. In fact, the average value may not explicitly exist within the data set at all. The average may be a number that never occurs but represents the data set's middle value. National culture is similar that it represents the overall population culture but may never exactly represent any specific organization or individual's actual culture.

Other national culture applicability challenges focus on the aspect that culture should not be viewed as a fixed notion but one based on what is constructed through interactions. That is, rather than view culture as something which is rigid and unyielding the practitioner or researcher should view culture as the outcome of interactions between the associated entities and something which continues to change. These cultural changes are an outcome of actions which occur within the group (Smits 2013, 30).

The view that a multinational team's culture is dynamic is in congruence with this book's theme that culture is an open system. As such, the team's culture receives and returns external environment energy. It is not a stagnant or a closed system but one which is dynamic and subject to change as the energy flows into and out of the team environment. To this end, the following sections highlight the various literature concepts which identify methods on how team culture evolves, adapts, or changes to fill the team's needs within the specific environment.

6.4 MANY TERMS SUCH AS CROSSVERGENCE, HYBRIDIZATION, MULTICULTURAL, CROSS-CULTURAL, CONVERGENCE, AND DIVERGENCE: WHAT IS WHAT IN THE WORLD OF MULTIPLE CULTURE MODELS?

The preceding section provided a foundational definition of a multinational project team. It also set the stage for discussing how a homogeneous team culture may actually be a rarity in any country, state, and potentially any project team. When looking at the organizational, team, or individual cultural level there is a distinct possibility that one will identify different cultural variations than what may be assumed when viewing any of these entities from the overarching national culture point of view.

Depending, as will be discussed in the next paragraphs, on the practitioner's or researcher's point of view, multinational culture interactions are influenced in different ways. These influences result in a modifying effect on each of the national cultures as experienced by that specific group and as an interaction of the various cultures. As the potential exists for many different cultures to reside, within a single country, the literature refers to the blending or merging of these cultures through various actions such as:

- Crossvergence
- Hybridization
- Multicultural
- Cross-cultural
- Convergence
- Divergence

The following paragraphs are intended to provide the reader a general overview of these various terms, theories, and research concepts which surround the multinational project team environment.

When two or more cultures are brought into close contact, a fusion of cultures can occur. One description of this fusion process is a crossvergence process (Jacob 2005, 522). "The term 'crossvergence' was coined by Ralston and colleagues in … [a] 1993 JIBS [Journal of International Business Studies] article, 'Differences in Managerial Values: A study of U.S., Hong Kong and PRC Managers'" (Ralston 2008, 27). The concept of a crossvergence culture is based on the idea that within the team members' interactions there is a combination of events and influences which occur at the social and business environment levels which ultimately impacts the team culture. As an example, some researchers identify that there is a direct linkage between society's value system and its technology. Within this concept, the literature postulates that technology is the catalyst which brings together society's cultural bonding more so than other sociocultural influences (28). These technology and social interaction events and influences become the driving forces behind the development of a new culture. It is this interplay of external and internal forces which shapes the team's resulting culture. The technology forces are also those items which will continue to mold the team's culture as it moves forward in time.

Crossvergence is inclusive of three categories: conforming crossvergence, static crossvergence, and deviating crossvergence. Conforming crossvergence is the process of cultural difference minimization over time.

As the team members interact, the different cultural underpinnings are modified such that a common culture emerges which will include attributes from each culture as well as new attributes. Static convergence is the state where the team's internal values may change over time, yet the value differences between the teams do not change. Finally, deviating crossvergence is the opposite of conforming crossvergence. That is, over time, rather than a common value system emerging between the associated groups, their respective cultural value differences actually increase. What this implies is that a greater difference in values occurs as time moves forward rather than what existed in the beginning (Ralston 2008, 37). Which crossvergence is in play, as noted in the descriptions, will be a factor of the overall environment influences, open system interactions, as well as internal and external forces.

The term culture hybridization and the related term culture hybridity are described as the cross between different cultures. These terms originate from "… biology and botany where it designates a crossing between two species by cross-pollination that gives birth to a third 'Hybrid' species" (Guignery, Pesso-Miquel, and Specq 2011, 2). The association or theory of culture hybridity is generally attributed to Homi Bhabha in his early 1990s works about postcolonial discourses. Specifically, Bhabha was instrumental in introducing the biological hybrid concept within the social cultural realm. His efforts were based within the postcolonial research on linguistic, political, and ethnic intermixing (3).

Hybridity literature is rife with challenges and attacks. As an example, hybridity is identified as a means of forced cooperation by the true power holders rather than a culture transformation. In this case, the power holder's primary objective is to reduce or eliminate differences for the achievement of their ultimate objectives rather than cultural change within the group based on cultural interactions (Kraidy 2002, 322). Another challenge is that hybridity is theoretically useless as all cultures are hybrids. That is, culture occurs through the interaction of individuals such that the resulting culture will always be a hybrid. As such, the literature challenges point out that either all cultures are hybridizations or discussions about culture hybridization are moot.

While many culture hybridization discussions are available, the application of culture hybridity, within project management, is not widely studied. Out of the limited project management literature resources comes the concept of think globally–act locally. What this implies is that the individual team members maintain the essence of two cultures. On the one hand, they have their national culture, which is the global view through which they normally interact with those external to the project team such as their families, friends, and others whom they deal with external to the project environment. The second culture is the hybrid project team culture which is a merger of the project team members' cultures into a functional project team culture, that is, think locally. In this example, the project team culture is created when the members merge or cross-pollinate their different communication styles, values, norms, and ethics into a new form which takes from each of the interacting cultures to form a hybrid between them. Further research is required in this area to provide additional data and information as to its applicability, or not, within the project team environment.

The terms multicultural and multiculturalism are some of the oldest referenced terms in culture research. These terms are the most basic forms and simply refer to entities which include people of different cultures. Historically multiculturalism has existed and been discussed since antiquity. As an example, "Professor Lawrence suggested that the 11th-century Turk al-Biruni, who learned Arabic, Persian, Greek, and Sanskrit and wrote 125 books in Persian and Arabic, might be a good symbol of Islamic area multiculturalism" (Gress 1999).

Multiculturalism projects have correspondingly existed virtually from the initial acknowledgment that project management is a unique management discipline. From the very beginning, project teams have existed that included participants from different cultures, and therefore, are multicultural.

As such, multiculturalism is generally used as a descriptive term which identifies the presence of multiple cultures within the environment rather than a process term in how different cultures blend together into a new or different culture. Within the project management literature, this is generally how the term is applied as well.

Multiculturalism is also applied as a normative term which refers to ideologies or policies. For the project management discipline, this could be viewed as the team's policies associated with acceptance and promotion of diversity within the team. That is, the team not only embraces the diversity of multiple cultures but encourages the inclusion of different cultures within the project team environment. Multiculturalism in and of itself does not explain or provide a deeper understanding on how the multinational project team should and can develop a cohesive and inclusive working culture which encompasses the diversity of the multicultural environment. Rather, it describes the general environment.

The phrase cross-cultural is broadly used in the descriptive and normative terms similar to multiculturalism. From a descriptive perspective, cross-cultural is used to describe two or more cultures which are present within a specific environment. Within the project management discipline, an example of this is the description of a project team which includes people with different cultures as a cross-cultural project team. It is descriptive and lacks an action component. Classifying a project team as cross-cultural does not require or drive the project manager or team members to take any specific action.

Cross-cultural is also generally applied as a reference to a specific type of research. Specifically, cross-cultural research involves comparing and contrasting different cultures to identify similarities, differences, and ultimately define new theories or universal cross-culture concepts. This approach is associated with knowledge expansion and development of tools, methods, and techniques on how the project manager and project team members can overcome the challenges of a project team which includes different cultures. Cross-cultural, as a research term is critical to the project manager and project team as research provides the data and information which can assist them in developing and coexisting within a cross-cultural environment as effective and efficient team members.

In summary, the term cross-cultural is generally applied in a descriptive manner to highlight or identify a project team which includes people from different cultures. As previously noted, this is descriptive and does not generate any action, need, or requirement. When applied in the context of research, cross-culture refers to developing a better understanding of different cultures through various compare and contrast approaches. The outcome of cross-cultural research is knowledge enlargement within cross-cultural environments. The research results provide input to the project manager and project team but do not drive action on how the cross-cultural team may develop a unique project culture or how to reduce the potential conflicts and risks which may occur.

The next cultural phrase is cultural convergence. Culture convergence, as with many cultural terms, has its origins outside of management research but it has been adopted by management disciplines. Cultural convergence refers to the process of different cultures coming together as a single entity. This is based on convergence theory where some of the earliest references occurred during the 18th-century in the field of economics.

Convergence theory postulates that societies start to obtain similar cultures as their industrialization state improves (Crossman 2014). Within projects, convergence theory would identify that the project team interactions result in a common culture and understanding convergence. In this context, the project team members embrace the merging of cultures, rather than hinder or resist the change (Alderman and Ivory 2011, 18). Stated slightly differently, the project team would not resist the culture convergence, so the resulting unique project culture would occur sooner and with a less negative impact. The change process is shorter, more efficient, and has greater positive impacts.

The converse to convergence is divergence theory. Divergence theory postulates that rather than a convergence of project team members' cultures toward a common point, they actually diverge. That is, within the divergent project team there is a lack of awareness, understanding, and acceptance of the legitimacy of the other cultures (Alderman and Ivory 2011, 18). In this situation, the team members fail to converge toward a common culture point. Rather the team experiences an increase in cultural differences with a corresponding increase in team interaction misunderstandings and miscommunication which ultimately reduces the project team's effectiveness and efficiencies. Divergence may occur as a result of different drivers such as perceived sense of right and wrong, individual priorities, perceptions of what are correct practices and communication processes (Yan 2009, 702). Divergence is disruptive and fraught with danger to the project team.

Divergence disruption and dangers are real risks to the project team's success capability. If the team members' cultures diverge such that a common set of values, practices, priorities, and norms cannot be established, the project team's effectiveness and efficiency will collapse. The results will potentially prevent the project team from being successful.

In summary, this section provides the reader a general overview of the various terminologies and theories associated with cross-cultural research and potential impacts that may occur within project teams. The intent of this section is to provide the reader a foundation on what the terms refer to and the various theories and concepts associated with them.

6.5   WHY BE CONCERNED ABOUT MULTINATIONAL PROJECT TEAMS?

The previous discussion provides a natural lead-in to the following questions. What do we mean when we refer to a multinational project team environment? What are the challenges and potential issues associated within a cross-culture environment as it applies to the multinational project team? How can the practitioner and researcher use this information within their areas of application? The following is intended to provide insight and suggestions to address each of these questions.

This section of the chapter looks at what cross-culture really means within the project team environment. As a side note, the following discussion uses the phrases cross-culture, cross-cultural, and multinational in the descriptive form. The use of these terms is intended to identify a project team environment which includes two or more cultures rather than terms intended to identify or describe specific research types, research methodologies, or specific project team cultural theories.

As stated in Chapter 2, the need to understand multinational project team interactions is an ever increasing necessity. In this era, it is almost redundant to say that the world is shrinking, globalization is occurring, or that the world is flat. Throughout the literature, one consistently encounters these terms and reference to globalization. This is a direct result of the dramatically increasing global financial interactions, knowledge sharing, commerce exchange, and social interactions which are the new norm in this society ("When Did Globalisation Start?" 2014). Over the last 20 years, the rate of international interaction with trade organizations, political entities, and social groups has dramatically increased when compared to all previous eras. One only has to walk through any store and look at the origin of the goods and products. A brief stroll through many retail stores will reveal a host of countries as the goods and product sources. Retail environments are very global.

An outcome of this worldwide interaction is a major impact on the various world cultures. The exchange of ideas, greater communications, and interactions expose a broader set of individuals to different cultures and the need to work within them (World Forum 2000). These cultural interactions impact each nation and result in change. Just compare the differences between East and West Berlin cultures before and after the Berlin Wall fell or the global financial impact that China's economy has as but two examples that clearly demonstrate the impacts globalization has on the world and national cultures as well.

Within this expanding globalization, the application of project management as a discipline, and development and utilization of multinational projects to achieve cross-border business objectives have correspondingly increased as well. Whereas early project management days existed more often within a homogeneous cultural environment, in the 21st century, multinational projects are extremely common and continue to increase on an annual basis. It is estimated that between 2010 and 2020, an additional 15.7 million project roles will be required (PMI 2013b). A large percentage of these new roles will be involved in global projects and involve multinational project team environments.

The influences of globalization, that is, increased global knowledge exchange, financial trade, and capital exchange are the driving forces for the increasing number of multinational projects around the world.

These multinational projects are in themselves driving forces for research as a means of understanding how the project team interacts within a cross-cultural environment. *Merriam-Webster* defines cross-culture as "relating to or involving two or more different cultures or countries" (*Merriam-Webster* 2014). This book modifies this definition to be in alignment with the definition of a multinational project team as relating to or involving two or more different cultures regardless of country or state origin.

So, what does globalization mean as it applies to the project management discipline or to restate the earlier question, what does it mean when we refer to the project team context as a multinational project team environment? The answer is centered within the global project team composition. Global project teams are inclusive of members from different nations which all have their own cultures. As such, this results in a cross-cultural project team that includes people from different nations. This is in alignment with this text's definition of a multinational project team and interaction of different cultures. The same is true of any multinational team, be it a company safety team, engineering team, or human resource team as but a few examples.

Within the research literature, there are two primary streams of thought as to the impact of multiple cultures within a project team structure. One stream of thought is that with greater cultural diversity the project team brings together a broader set of skills, tools, and knowledge than which is present in a homogeneous or single culture based project team. This stream is supported by research which identifies the aspect that each culture brings with it different views, skill sets, knowledge bases, and processes. It is these differences which provide the multinational project team the ability to see things in different views and to create unique solutions (Ochieng and Price 2010, 449). This position is grounded in the concept that different cultures provide a broader base of potential insights, perspectives, and generation of alternatives specifically due to the different cultures' world views.

The other stream of thought is that multinational project teams complicate the team members' interactions and relationships (Haas and Nüesch 2013, 5). This complicated relational attribute refers to how the team members communicate, how they make decisions, what are accepted values, as well as accepted norms of action within the project team. An example of supporting research is found in the discussion on project success. As noted throughout the project literature, project success is a difficult metric to achieve on a consistent basis. There are many causes for this challenge, yet these challenges increase when the project environment exists within a multinational project team context.

Within the multinational project team environment, the team members not only face the *normal* range of project success challenges but also they face different cultures. These different cultures may have varying values, needs, decision-making processes, norms, and values (Ochieng and Price 2010, 450). In this situation, the different cultures are viewed as potentially conflicting and disruptive versus a source of expanded knowledge base, different approaches to a common problem, and different but agreeable ideas.

Therefore, to answer the first question, multinational project teams include more than one culture which gives rise to the potential that greater skill sets and alternative generation exist along with cultural-based relationship complications. This, coupled with ever increasing globalization and a corresponding increase in multinational project teams, highlights and strengthens the need to develop and enhance cross-cultural project environment skills, tools, and knowledge. Multinational team interaction and subsequent advancement in cross-cultural capabilities will result in improved working environments that can enhance the potential outcome of a successful project.

The second question to address is: What are the challenges and potential issues associated with cross-culture research? To start this discussion, a review of cross-cultural research is in order. Cross-cultural research has its foundation in the 19th-century work of anthropologists Edward Burnett Taylor and Lewis H. Morgan. From these early efforts cross-culture research has expanded into virtually every area from psychology (Toledo 2014), management, and project management.

The project management discipline is a late comer to the study of cross-cultural research by the very fact that project management is a recent discipline in and of itself. As such, project management cross-culture research is not totally lacking but the breadth and depth of it needs further attention and focus.

Predominately due to the recent advancement of project management as a specific discipline as well as the difficulty and cost of doing cross-cultural research, there is an unfortunate lack of definitive project management culture research. As such, this limits the availability of research results, tools, and methods which the interested cross-culture project management professional may draw upon. This is supported by the project management research field which identifies the minimization of cross-cultural project management data and information while acknowledging that culture is one of the areas that all projects have in common and that culture can either support or negate project success (Henrie and Sousa-Poza 2005, 6).

Expanding this to cross-cultural project environments, there is even less data and information available to the project management professional but it is a growing area of need and concern in alignment with the ever increasing globalization and expanding multinational projects.

The final question to be addressed is: How can the student, practitioner, and researcher use this information within their areas of study and application? The answer is based on the concept that culture has an impact on all projects. The potential resulting impacts are magnified when there are multiple cultures within a single project team. If the project manager does not manage the cultural issues and needs successfully there is a significant probability that the project success is placed in doubt, team members' satisfaction level will be negatively impacted, and it becomes highly unlikely that an effective and efficient project can be achieved (Zwikael, Shimizu, and Globerson 2005, 454).

Therefore, it is imperative that cross-cultural project professionals develop an extensive understanding of cross-cultural aspects and project management skill sets. When working within multinational project environments, it is essential that the team members have these skill sets and a cross-culture knowledge base to establish the foundation for a successful project. This is because many project failures are not driven by technology, desire, policies, procedures, or willingness to participate but by people issues which the project management team either does not address or ineffectively addresses (Gunding 2013, xi).

Clearly cross-culture research and knowledge is critically important. Yet, from the project management discipline's point of view, there is a distinct shortage in data and information which is specifically focused on the challenges within a project team environment. The literature identifies that researchers have been and continue to be focused on analyzing and attributing project team cultural issues at the national level when the issue is at the team level. Therefore, focusing nationally, rather than at the team level, provides minimal to no benefit within the project team environment (Chevrier 2003, 142).

This statement highlights the aspect that there is not a universal culture or a universal approach to working with people from different nations within a single team environment. The uniqueness of each national culture requires the project management practitioner to develop an understanding of that specific culture's unique attributes and how it may conflict with the other cultures which exist within their project environment.

This then becomes the answer to the last question. That is, developing a deeper and better understanding of different cultures and their potential interaction effects provides practitioners expanding skill sets which ultimately contributes to being successful within their areas of expertise. Applied knowledge is essential to be successful.

To this end, the following sections provide the reader a general overview of cultural attributes and potential impacts within cross-cultural environments. The intent of the following discussion is to provide the reader a deeper understanding of how different cultural attributes may interact within a multinational environment and ideas on how to work within that context.

6.6 COMMUNICATIONS: SO ESSENTIAL BUT SO EASY TO MISUNDERSTAND

Cultural interactions, be it homogeneous or cross-cultural, involve people interacting with other people. Understanding how these interactions occur is a cultural and cross-cultural environment focus. Or stated slightly differently, understanding these different interactions is based on studying how individuals interact, make decisions, how conflicting values and norms are handled, and communications occur (Adler 1983, 226).

Regardless of whether the context is culture or cross-culture based, the common reference subject is the individual and his or her culture. When a project team consists of two or more nationalities, this establishes a cross-cultural context where one literature trend states that communication can be influenced by the individuals' values and norms as influenced by their power distance culture attribute (Muller and Turner 2004).

Effective and efficient cross-cultural communication is an essential attribute if the project team is going to have a chance to be successful. If the project team continuously experiences communication misunderstandings, communication gaps, and communication lapses there is a slim chance that the project will go as planned. Communication misunderstandings can happen for any number of reasons such as conflict between cross-culture team members which may include any number of reasons such as religion, varying personal beliefs, different values and norms, formal or informal interactions, as well as family and friend obligations (CSU 2014). Just think about recent issues you have had with your significant other, family member, team member, and so forth. Very often the root cause of these issues is grounded in failed communications. You might have communicated too much information or not enough, your intended message was not received in the intended form, or your misunderstood the message sent, and so on.

While close, personal miscommunications are not rare it just underpins that the potential for cross-cultural communications is even higher. These types of communications, that is, cross-cultural, issues are grounded in the fact that people of different nations have developed different communication styles and processes which include many aspects of the environment such as the way they communicate within a hierarchical class system and corporate structure. One way to refer to this communication is by the culture power distance attribute.

Power distance refers to the fact that within all culture environments different individuals have different levels or unequal levels of power. Specific to the corporation there is a hierarchical power structure which starts at the chief executive level and progresses downward to the daily worker who may be an engineer, team assistant, or the janitor. Within the project team, the power structure is envisioned as starting with the project manager, proceeding to individual team leads; such as the engineering team lead, construction team lead, and procurement team lead; then down to the individual team members such as the engineer, construction team member, and procurement specialist. Each level has a specific set of responsibilities and associated powers. An example is that the project manager has the power to approve all procurement requests while the procurement specialist's power is limited to the identification of and preparation of the recommended procurement requests. Procurement specialists do the research, analysis, and request generation. They are not authorized to approve the procurement.

Developing an understanding of how the individual's power distance culture interacts within cross-culture project team communications is essential. This statement is grounded in the fact that as the individual's power distance attribute helps define the way the team members communicate with their superiors, peers, and subordinates it will correspondingly have an impact on the project. If the project manager is not aware of various team members' power distance attributes, this can set the stage for communication misunderstanding. Reducing the negative impact of power distance enhances how the project team communicates which contributes to improved performance as well as increasing the probability of project success. The literature is clear that communication is a critical project implementation success factor (Pinto and Slevin 1989) and is positively related to improved project performance (Rodwell, Kienzle, and Shadur 1998). Communication has been identified in all personal interactions as not only essential but vital. Poor communication can generate significant disturbances and the individual's potential communication success is grounded within the framework of the individual's power distance cultural attribute (Muller and Turner 2004, 407).

The following examples are provided as a means to assist in developing an understanding of how cultural power distance may be demonstrated within a project environment. The first example includes a project team where all members have a similarly low power distance score such as what may be found in Austria (11) and Denmark (18). We must also assume that the team member's individual power distance score reflects that of the nation he or she is from. A cautionary side note is in order. That is, one cannot automatically assume that an individual from any nation will fully reflect the national culture from which they are from. National culture indexes and descriptions reflect the general cultural attribute of the nation not of an individual.

In this example, the low power distance score identifies that the team members view communication in a consultative or democratic way. In this context, there is less command and control function where the superior directs the subordinate to do specific things and the superior does not expect or is surprised if the subordinate questions the direction, provides input, or offers alternative suggestions. In the low power distance project team environment, communication is bidirectional and team members are very comfortable with questions, challenges, suggestions, and collaboration.

A second example is where the project manager has a low power distance score, such as one from Denmark, while the majority of the team members have a high power distance score, such as Guatemala with a score of 95. If the project manager is not aware of the Guatemalan team members' high power distance, he or she would not be aware of the difference in culturally accepted communications. In this situation, a cultural-based communication misunderstanding is fully possible. That is, the project manager is culturally used to and expects the project team to provide feedback, suggestions, ideas, and even challenge the offered decisions. Conversely, the project team members' culture-based communication style is to accept the direction as provided with no bidirectional interchange. That is, they are not used to or comfortable entering into a dialogue with their superior when given an order. As such, there is a real potential for communication misunderstandings within this situation in that the project team will do exactly as directed. This response is culturally based, regardless of whether there is a better way of doing the assigned task, they have different ideas on how to proceed, or even if they do not agree with the directions received. As such, the project manager may become frustrated with the lack of project team communication and the project team may become frustrated with the project manager for providing directions which are clearly wrong or where there are better ways to accomplish the intended results.

The third example is the situation where the project manager has a high power distance communication style while the project team has a very low one. In this situation, the project manager is not accustomed to their subordinates asking questions, providing suggestions, and even challenging the decision. As a project manager with a high power distance communication style, he or she expects the team member to do as directed without a dialogue. This scenario fosters a condition where the project manager is frustrated that the project team is not just doing as they are directed while the project team members are equally frustrated in that they feel that the project manager does not value their input.

Potential communication issues transcend beyond the individual's power distance attribute to include the accepted language medium. In various technical fields, English has been adopted as the *standard* language when dealing within multinational environments, be it published literature, conference presentations, or project team environments. It is often very common to find bilingual project team members who converse freely in English as well as their national language. Yet, just because a person speaks English does not eliminate potential communication issues which are associated with the communication medium. The literature is clear that just because someone speaks English does not mean that he or she understands English as spoken by a native English speaker. There are several aspects of this communication challenge such as language-specific innuendos, doublespeak, slang, and common versus formal speech patterns (Schermerhorn and Bond 1997, 188).

Therefore, for the project manager and the project team a key element to understanding the project team's culture-based communication is through education, observation, and experience. From the educational perspective, the project manager and even the project team members should study various literature sources as to how specific project team cultures generally view power distance and what is seen as acceptable communication styles. Education, training, and experience provide the project manager and project team members the ability to enhance the project team's capabilities in the complex multinational communication area (Schermerhorn and Bond 1997, 191).

The key elements which the practitioner should take away from this section are that:

- Communication is culture based.
- Power distance culture attribute impacts how communications occur.
- Effective communications are an essential element in any project success.

- Poor communication has been shown to prevent effective team interaction.
- Project personnel must be aware of and have an education in the various cultures which exist within the team.
- Observing, learning, and adapting how to communicate within the cross-culture team are required by all.

It is one of the project manager's responsibilities to ensure effective communication occurs. This is accomplished through culture-based educated leadership.

6.7 INDIVIDUALISM OR COLLECTIVISM: ARE WE INDIVIDUALLY ORIENTED OR GROUP ORIENTED?

In this section, the concept of how the team interacts within itself is explored. Specifically, this chapter explores the concept that the project team performs their work as a set of individuals or whether they work together as a collective project team society. The difference between individualism and collectivism, as is discussed further in the following paragraphs, is culturally based and impacts the project environment.

Cross-cultural research has clearly identified that different national cultures view the importance of the individual versus the importance of the group differently. Some cultures view the individual and his or her personal interest above that of the collective group. These societies are described as individualist. In these societies, the individual is paramount and everyone is expected to be self-fulfilling and looking after his or her own. People within the individualist society tend to have a small set of very close relationships and tend to focus on the well-being of the individual before that of the broader collective group, such as the project team, and drawing individual attention to oneself is a positive trait (Hofstede 1997, 51).

The converse of individualism is the collective society. In these societies, it is paramount to belong to and support the group. The group and its success is the primary focus of all who are associated with it. Drawing individual attention to oneself and maximizing the individual's position over that of the collective society are generally not accepted (Hofstede 1997, 51). As noted within the cultural literature, in collective societies, the group's needs take precedence over the individual. In extreme cases, individual heroes tend to not occur as all efforts are focused toward the group and not the individual. It is better to sacrifice the individual advancement or individual success if in the end the collective group receives the praise or benefit rather than the person.

Understanding if the project team consists of team members from an individualist society or collective society is essential from several perspectives. First, understanding each team member's orientation is essential for proper job assignment. As an example, if the project required a collaborative and cooperative group effort, assigning a set of individualistic-based team members to that activity would probably not produce the best results. In this situation, the individuals, as is their culture, would be working toward their own best interest which may not be associated with providing the optimum collective solution. That is, the individuals may become a roadblock for the collective team's successful solution.

The converse is also valid that assigning a culturally based collective person to a highly individualized task may not produce on optimum outcome within the concept of time and personal satisfaction. That is, if a project activity required an individual to take full responsibility to produce the result in a singular mode, this is not the optimum situation for a collective-oriented team member. In this situation, the team member would be forced to work in an isolated environment where success or failure is viewed within the context of the individual, not the group or team. This is counter to the collective culture which strives to support the group and not the individual. This situation could generate conflict within the assigned person which may prevent him or her from performing the assigned task to the level required.

As noted in the research literature:

> The amount of interaction the job requires of subordinates is an important consideration for managers in analyzing their work environment. ... [Where] cooperation and teamwork [are] required not only among the workers but also between the workers and their superiors ... workers preferred and worked better under employee-centered supervisors.... [While, where the task] did not depend on others for accomplishing their task ... [the workers] preferred task-oriented supervisors. (Hersey and Blanchard 1982, 137)

Understanding the team member's individual versus collective culture attribute is an essential skill and knowledge set. It provides invaluable assistance in job assignment as well as identifying how the project manager should approach oversight of the team member's work.

Another aspect of understanding the team member's individualism or collectivism culture trait involves the project team member's satisfaction level. While the project team's primary focus is to complete the agreed on project, of equal consideration should be the team member's project participating satisfaction value. Research clearly identifies that there is a relationship between the team member satisfaction level and performance and the team's collective satisfaction and performance metrics. Satisfied team members, and subsequently a satisfied team, have higher levels of performance as well (Zeitun, Abdulgader, and Alshare 2013, 292). Project team members with higher satisfaction levels exhibit increased performance at the same time. Higher satisfied team members contribute to higher performing teams which correspondingly cascades into improved probability of project success. The converse is also valid in that project teams with low satisfaction level team members negatively impact the overall team satisfaction which ultimately results in lower project performance, and the probability of project success also decreases. Happy people create a positive atmosphere which is a positive influence while unhappy people contribute to a negative atmosphere. A less than desirable positive atmosphere cascades into the overall project and results in a higher probability that success will not be achievable.

As noted earlier, if the project manager understands the individual team member's individual versus collective culture attributes they can assign the team members to activities accordingly. This supports the project in achieving its end goals as well as providing the team members a supporting project environment which can enhance their satisfaction level.

A third aspect of the individualism and collectivism culture attribute, within the project team, is the concept of conflict. As in all team activities, be it a marriage, a long-term personal relationship, working relationship, or project team interaction the potential for conflict is present. Within the project environment, conflicts can and do occur in relationship to resource assignments, policies, procedures, values, norms, and decision-making processes. While homogeneous project teams can and do have a potential for conflict, within a cross-culture environment the potential increases. Regardless of homogeneous or cross-cultural project team environments, it is clear that conflict does not enhance a project team's performance or overall satisfaction either (de Dreu and Weingart 2003, 741).

An example of how cross-cultural project team interaction can result in conflicts is where culturally grounded individualist team members and collective team members must work within a common area toward the project's final objective. Each team member would have his or her primary focus on his or her own world view, as to how the team should interact, not on how the team can collectively be successful. This sets the stage for conflict along many lines. One potential conflict can occur as the individual centered team member would probably not want to share success or work activities with the collective team member. Again, this is driven by the individual need of looking out for himself or herself. Concurrently, the collective-oriented team member may become disillusioned or upset with the individual-focused team member for *grandstanding* and taking all the credit when the collective team was the environment which fostered the project success. These are different world views and sources for conflicts. Project managers must be aware of and have developed strong methods and approaches to handling such situations.

6.8 PROJECTS ARE RISKY BUSINESS: WHAT IS YOUR PROJECT TEAM MEMBERS' UNCERTAINTY QUOTIENT?

By definition, projects are risky business endeavors rife with uncertainty. These endeavors are implementing something that is unique, within a temporary project team structure, which is encompassed within a limited time window, restricted budgets, and quality constrained environments. At the same time, the project team consists of individuals who probably have different risk tolerance levels and uncertainty anxiety levels. One way to measure the team members' anxiety tolerance is within the culture attribute of uncertainty avoidance.

Regardless of where you go in the world, what organization you may work for, or what project team you will be on, handling uncertainty is "… part and parcel of any human institution in any country" (Hofstede 1997, 110). In all activities, there is a level of uncertainty. The more unique the undertaking, the greater the uncertainty which may exist. Within all environments is the corollary of uncertainty avoidance. Uncertainty avoidance is one means of describing and to a certain degree measuring to what degree a team member feels threatened within unpredictable situations (113).

As an example, if a team member becomes agitated, frustrated, or afraid during uncertain times he or she has a low uncertainty avoidance quotient. On the other hand, if the team members accept encountered uncertainty as a normal state of the project without hesitation or fear they have a high or acceptance uncertainty level. That is, the unknown is not scary or a source of negative feelings.

Within the project research literature, there are three themes associated with uncertainty. The first theme is that uncertainty exists in all projects and that it must be managed. Uncertainty management involves various tools and techniques such as development of plans, procedures, and schedules with appropriate tolerance and budget contingencies. Lower uncertainty quotient team members will generally view development and implementation of these tools in a positive light. Conversely, higher anxiety quotient level team members may view all the tool planning and implementation as constraining, as an excessive expenditure of time and resources, or both.

The second theme is that uncertainty management is essential to project success and team satisfaction. Uncertainty management requires that the project team have a plan on how to manage it. If the project team has not established the uncertainty management plan, with associated procedures, and is following that plan a reactive, not predicative, project environment ensues. Rather than implementing a project plan, the project team will be reacting to uncertainty events and putting out *fires*. This is a reactionary, rather than planned action oriented, project uncertainty environment which often leads to lower project success and team satisfaction. Team members with lower uncertainty quotients tend to exhibit higher levels of anxiety, discomfort, or agitation when the project is reacting versus planned implementation.

The third theme is that culture is a key factor in how uncertainty is approached and managed. As identified in research, there is a relationship between the organization and team cultures and how effective is the required uncertainty management (Karlsen 2011, 241). This can be extrapolated to the fact that the project team culture will have a significant influence on how effective the project team uncertainty management processes will be. When this factor is applied to the cross-culture project team environment, the complexity of this process increases.

Cross-culture uncertainty project management processes may be considered of greater complexity when compared to a homogeneous project environment. This increased complexity quotient is derived from the interactions and conflicts which may occur between the different cultures' uncertainty avoidance attributes. In this environment, the project team may include people with low as well as high uncertainty attributes. Those team members from uncertainty avoiding cultures will be looking for policies, rules, and procedures to be in place and strictly enforced. These tools are used to reduce the project uncertainty which fits with uncertainty avoiding culture's desires and needs.

Conversely, project team members who have higher, that is, greater uncertainty avoidance acceptance avoidance attributes do not require or seek an extensive set of methods, tools, or approaches to reduce project uncertainty. This set of project team members is very comfortable in a higher uncertainty state and more willing to "go with the flow." These team members see the various uncertainty approaches as restrictive, hindering development of novel solutions, and counter to being an effective project team. Tight structure, mandatory procedures, and pinpoint-focused policies tend to restrict these individuals and ultimately reduce their overall project participation satisfaction.

What can be taken from the discussion on project team uncertainty is that when low and high uncertainty project team members work together, there is a strong potential for an increased overall team anxiety level. The team members may view the project environment as either too loose with insufficient structures, rules, and policies to be effective, or they will see it as being held within a straitjacket of rigid structures, rules, and policies which limit their ability to be effective.

This duality of the uncertainty situation is acknowledged by the project literature with a few examples or models on how the project manager should maneuver within the conflicting needs. There is a given in virtually all projects that uncertainty will exist. It is also a state which challenges each of the team members and an aspect of the project which the project managers are often significantly challenged in how to manage the situation (de Meyer, Loch, and Pich 2002, 60).

Within this difficult situation, there are a few approaches that the project manager can apply to reduce the overall team uncertainty. To start with, and as a common research theme, project managers must develop an understanding of their team members' cultural uncertainty index. This understanding must include adding to their knowledge of how low and high uncertainty cultures approach unique and unknown events. They must also understand how the range of uncertainty cultures can and do generate conflict. Project managers must also develop an understanding of what range of uncertainties may exist within their project environment. In an ideal state, they would understand each team member's uncertainty avoidance value and how it could be minimized within the context of the project activities. Realistically, project managers will rely on their direct reports and general observations to develop this understanding and to develop a corresponding management approach.

By increasing the overall environment uncertainty quotient knowledge, project managers are placed in a better position on how to address the potential conflicts. This knowledge will allow them the ability to leverage the information in many ways which are focused toward addressing the team members' needs as well as the project objectives. One potential approach would be to assign project team members to activities which fit within the uncertainty framework. Those with a need for more structure can be assigned to those activities which provide this, while allowing for the assignment of less structure required team members to areas that are more dynamic. This type of job assignment also assists by avoiding placement of diametrically opposed uncertainty resources within an activity which will generate an automatic conflict. An obvious caveat to this approach is that the required skill sets and assignments can be linked.

Understanding the project team's uncertainty value also provides the project manager guidance on the level of structure that will be required. If the team, overall, requires a lot of structure then the project manager can ensure that sufficient rules, policies, procedures, and requirements are in place to support this environment. Conversely, if the project team is more attuned to flexibility and dealing with anxiety then the project manager can ensure the right level of structure is in place to support the team's needs.

Uncertainty within the project environment is a given and how the project manager handles the conflicting cross-cultural anxiety needs is a challenge.

6.9   CROSS-CULTURAL ENVIRONMENTS: THE OTHER CHALLENGES

In the preceding pages, the cultural aspects of communications, individualism, collectivism, and uncertainty avoidance were discussed. Within the project, the project manager will also face many other multinational cultural aspects and potential issues. Some of these other cultural aspects include values and ethics. In looking around the world, one finds a range of values and ethics which are deemed culturally acceptable but diametrically opposed. Within multinational team environments it is highly probable that the team leader will experience or identify the various conflicting culturally acceptable aspects of values and ethical behavior. The following paragraphs briefly discuss these cultural attributes and how they are important to team managers such as the project manager.

As there is a potential that multinational teams may include people with different value systems, what does this mean to the project manager? First, one must understand that values are the very foundation of the individual and organizational character. An individual's core values direct and support every interaction and all decisions (Deal and Kennedy 1982, 21). Yet, every discussion on values must acknowledge that each culture has a range of values within it. Some of these values are viewed as important, essential, and virtually inviolable while others are not.

Within the team environment, the potential for conflict is highest between cultures where core and non-negotiable values clash. As an example, part of your project team's values are inviolable that women have the same rights and capabilities as men. Yet, part of the project team's values clearly reject the concept of women working within a male environment and having equal say and equal powers. In this situation, aspects of each individual's core socially acceptable behavior become challenged by an opposing culture. In this example, some of the project team members insist that inclusion of women is an absolute. Yet, other team members' cultural foundation or core social value rejects the concept that women can work outside the home and they do not work with men who are not of their family. Clearly, these opposing culture values are mutually exclusive and probably inviolable at the team member's level.

An approach to overcoming such conflicting core values predominately resides in the recruitment process. This resolution recognizes that the team leader or project manager will not be able to alter a team member's core values. As such, the team leader or project manager who is interviewing potential team members must clearly identify if insurmountable value conflicts may arise based on the prospective team members' core values. It is at this point where potential and volatile conflicts can be avoided rather than included in the team. It is a mistake to believe that a team member can alter, may desire to alter or change those core values. These values are derived at a very young age and are so tactic in nature that the individual cannot elucidate why the values exist, other than these are what he or she believes and are facts. Even if the team members can communicate why they hold the values they do, this does not imply that they can or would be willing to change these for the betterment of the project team.

Multinational team managers or project team managers may also face challenges associated with competing ethics within the project team. But what does competing ethics really mean?

Ethics and ethical behavior have been and are often described in terms such as doing what is right, following your religious convictions, not deviating from what feels right, or doing what is socially acceptable. Yet, all of these general descriptions lack in their inclusion of the broader society and fail to clearly say what ethics is. One ethical definition, which tries to overcome these types of general statements, is that "... ethics refers to well-founded standards of right and wrong that prescribe what humans ought to do, usually in terms of rights, obligations, benefits to society, fairness, or specific virtues ..." (Velasquez et al. 2014).

Around the world there are different acceptable ethics views. As an example, within one culture it is acceptable to be late or not come to work without any advance notice while in other cultures punctuality and reliability is a core ethical standard. Or people from one culture may think it is ethically correct for young children to be working in hazardous industries while people from other cultures disagree and even have laws against such practices. There is also the situation where it is not only ethically accepted but fully expected that you will award all work to your family and friends first before seeking any external source. This is in conflict with other cultures which view such practices as wrong and are not allowed. The point is that what is viewed as standards of right and wrong vary, or their ethics, depending on the person's culture. When cross-culture teams are formed, there exists the potential for this type of ethical conflict.

Within the realm of project team ethics literature there are few research articles, models, or approaches presented. This places the project manager in the position of extending beyond the project literature to look for ways of understanding and managing these types of ethical issues.

One approach to developing a deeper multinational team with conflicting ethics issues resides in leveraging other disciplines' research, training, and knowledge. As an example, from the management literature one decision-making ethical model is shaped within different contexts. The first context is one of moral awareness. This context identifies that the manager, in this book context the project manager, must have the ability to view, understand, and derive the situation within a pure moral view. Of course, project managers will have their own moral underpinnings to be aware of when viewing the broader project teams which may cloud their understanding of the moral contexts which are present.

The second context encompasses moral decision making. This context views the process or model of making a decision as a moral process more so than a logical process. What this implies is that within the decision-making environment you are in, the ultimate decision will always be viewed in the light of a morally right or wrong decision rather than a logical thought process which derived a conclusion based on amoral cognitive decisions. A major ethical problem faced by expatriate project managers, within this decision environment, is that they may have a significantly different moral position that they are unable to overcome.

That is, they may find themselves in a position where their morality-based decision-making processes are in opposition to others within the team.

The third decision-making context is associated with the ability to clearly identify moral intent. What this refers to is the need to prioritize moral values over other values that come into play. If you violate higher level core values to satisfy lower level values it can generate significant conflict both within the individual and within the team. As an example, it has come to your attention that your administrative aid has been completing and signing the time sheet of another individual. While the reported time is correct, it is indirect violation of corporate policy for anyone to sign a time sheet for another person. Your administrative aid is a single parent with many bills and several children to take care of. You are faced with the moral context of either enforcing company policy, with the full knowledge that no one obtained any financial benefit from the actions, and watching the major impact to this struggling family or looking the other way and having other team members become aware that you did not enforce the company policy.

An approach to handling these moral conflicts is referred to as developing a moral intent (Ho 2011, 517). Developing an understanding of the moral intent is a four-step approach which is intended to assist the decision maker in determining if an action is ethical or not and when to implement the action or not. In general, the four steps involve (1) identification of the ethical issues that are impacting or creating the dilemma, (2) identification of intentionality, that is, the processes, actions, or activities that you believe are morally available, (3) select the way forward which meets or fulfills your moral standards, and (4) reflect on the resulting actions. What you will notice is the final decision or action taken must be within your morally acceptable *window* as ultimately the decision is grounded on your values and as has been noted it is very difficult to violate these. While the process provides a foundation for analysis and action it also emphasizes the vagueness of determining what is ethical or not and stresses how the decision maker can be placed in a significantly conflicting position. That is, in any given situation the team member may view his or her action as fully moral while others will not. Your values and theirs are not in alignment or agreement and easy compromise is not achievable.

SUMMARY

In summary, this chapter provides the reader a background and foundation of various aspects which multinational (cross-cultural) teams may exhibit which predominately do not occur within teams that have a consistent or homogeneous culture base. To achieve this objective, the chapter briefly discussed the difference between multinational and cross-cultural teams. A takeaway from this discussion is that while each nation has its own culture, the entry of different nations' members into a joint project team results in a range of cross-cultural environments. Some of these environments are very similar while others will be found to be very different.

The chapter also discusses the various challenges that the project management practitioner or team leader faces within this context. The potential cross-cultural issues associated with team communications, power distance, individualism and collectivism, as well as uncertainty avoidance were discussed. The chapter also outlined the issues and limited research around cross-culture conflicting morals, values, and ethics.

At the end of the day, the team manager is the social architect for his or her cross-culture project team. It is imperative that he or she has extensive knowledge of the issues associated with this context as well as concepts, methods, and processes on how to obtain the most efficient and effective output of the team in a challenging, yet rewarding, environment.

Historically, the number of cross-cultural teams continues to increase year by year. The forecasts show this trend as continuing with an even greater need for culturally aware, culturally sensitive, and culturally attuned team professionals. This book was written with the intent to provide the reader a firmer foundation within the context of culture in homogeneous and cross-cultural teams. Projects and project management as well as team work environments can be fun, challenging, as well as providing opportunities to be involved in developing and deploying unique products and services in multinational environments. To be successful, understanding culture and knowing how to structure a team's culture is imperative for success.

In summary, an objective of this book is to assist the team leader and team members in increasing their understanding of team culture as well as increasing the understanding and awareness of the importance culture has within a team's environment. This book also provided the reader information on some tools, techniques, methods, and methodologies which can be leveraged to assist them in the challenging opportunities. By being proactive in increasing your knowledge and skills, you not only ultimately increase the probability of being a fully productive and successful team member but you also provide a foundation which can ultimately produce both team and personal satisfaction.

REFERENCES

Abrahams, P. 2013. http://www.brainyquote.com/quotes/keywords/culture_4.html (accessed August 3).

Aditya, R.N. 1999. "Definition and Interpretation in Cross-cultural Organizational Culture Research: Some Pointers from the GLOBE Research Program." http://prevetteresearch.net/wp-content/uploads/image/leadership/resources/Definition%20and%20Interpretation%20in%20Cross%20Cult%20Org%20Res.pdf (accessed August 4, 2013).

Adler, N.J. 1983. "Cross-cultural Management Research: The Ostrich and the Trend." *Academy of Management* 8, no. 2, pp. 226–32. doi: http://dx.doi.org/10.5465/amr.1983.4284725

Alderman, N., and C. Ivory. 2011. "Translation and Convergence in Projects: An Organizational Perspective on Project Success." *Project Management Journal* 42, no. 5, pp. 17–30. doi: http://dx.doi.org/10.1002/pmj.20261

Altinay, L. 2008. "The Relationship between an Entrepreneur's Culture and the Entrepreneurial Behavior of the Firm." *Journal of Small Business and Enterprise Development* 15, no. 1, 111–19.

APM (Association for Project Managers). 2013. http://www.apminfo.com (accessed October 8).

Archibald, R.D., I. DiFilippo, and D. Di Filippo. 2013. "The Six-Phase Comprehensive Project Life Cycle Model Including the Project Incubation/Feasibility Phase and the Post-Project Evaluation Phase." http://www.iil.com/downloads/Archibald_Di_Filippo_ComprehensivePLCModel_FINAL.pdf (accessed July 9, 2013).

Aronson, Z.H., A.J. Shenhar, and P. Patanakul. 2013. "Managing the Intangible Aspects of a Project: The Affect of Vision, Artifacts, and Leader Values on Project Spirit and Success in Technology-Driven Projects." *Project Management Journal* 44, no.1, 35–38. doi: http://dx.doi.org/10.2139/ssrn.2348759

ASEM (American Society of Engineering Management). 2013. http://www.asem.org/asemweb-emj.html (accessed October 25).

Auch, F., and H. Smyth. 2010. "The Cultural Heterogeny of Project Firms and Project Teams." *International Journal of Managing Projects in Business* 3, no. 3, pp. 443–61. doi: http://dx.doi.org/10.1108/17538371011056075

Axelrod, R., and M.D. Cohen. 2000. *Harnessing Complexity: Organizational Implications of a Scientific Frontier*. New York: Basic Books.

Becker, K.H. 2003. "Improving International Project Success." *The Journal of Technology Studies* 29, no. 1–2, pp. 51–55.

Bellou, V. 2008. "Identifying Organizational Culture and Subcultures within Greek Public Hospitals." *The Journal of Health Organization and Management* 22, no. 5, pp. 496–509. doi: http://dx.doi.org/10.1108/14777260810898714

Berger, A. May 2000. "The Meanings of Culture." *M/C Journal* 3, no. 2. ISSN 1441-2616.

Bertalanffy, L.von. 1969. *General System Theory: Foundations, Development, Applications.* New York: George Braziller.

Boehm, B.W. May 1998. "A Spiral Model of Software Development and Enhancement." *Computer* 21, no. 5, pp. 61–72. doi: http://dx.doi.org/10.1109/2.59

Booz & Co. 2013. *Measuring and Analyzing Corporate Values during Major Transformations.* http://www.booz.com/media/uploads/MeasuringandAnalyzingCorporateValues.pdf (accessed September 3).

Brady, W.H., and S. Haley. 2013. "Storytelling Defines Your Organizational Culture." *Physician Executive* 39, no. 1, pp. 40–43.

Bredillet, C.N. August 2011. "From the Editor." *Project Management Journal* 42, no. 4, pp. 2–3. doi: 10.1002/pmj.20257

Bredillet, C.N., F. Yatim, and P. Ruiz. 2009. "Project Management Deployment: The Role of Cultural Factors." *International Journal of Project Management* 28, no. 2, pp. 183–93. doi: http://dx.doi.org/10.1016/j.ijproman.2009.10.007

Brewer, P., and S. Venaik. 2012. "On the Misuse of National Culture Dimensions." *International Marketing Review* 29, no. 6, pp. 673–83. doi: http://dx.doi.org/10.1108/02651331211277991

Brookes, N., and R. Clark. May 1–4, 2009. "Using Maturity Models to Improve Project Management Practice." Paper Presented at POMS 20th Annual Conference, Orlando, FL.

Bushell, S. September, 2004. "Taking Ad Hoc Teams to Task." *CIO* 12, no. 43.

Carayannis, E.G., K. Young-Hoon, and F.T. Anbari. 2003. *The Story of Managing Projects: An Interdisciplinary Approach.* Westport, CT: Praeger Publishers.

Caupin, G., H. Knoepfel, G. Koch, K. Pannenbacker, F. Perez-Polo, and Chris Seabury. June 2006. "IPMA Competence Baseline." Version 3.0. International Project Management Association.

Cherry, K. 2013. "Introduction to Research Method." http://psychology.about.com/od/researchmethods/ss/expdesintro_2.htm (accessed September 6).

Chevrier, S. 2003. "Cross-cultural Management in Multinational Project Groups." *Journal of World Business* 38, no. 2, pp. 141–49. doi: http://dx.doi.org/10.1016/s1090-9516(03)00007-5

Christopher, W.F. 2007. *Holistic Management: Managing What Matters for Company Success.* Hoboken, Hudson: John Wiley & Sons.

Cleland, D.I., and L.R. Ireland. 2002. *Project Management: Strategic Design and Implementation.* New York: McGraw-Hill.

Cleland, D.I., and W.R. King. 1983. *Systems Analysis and Project Management.* 3rd ed. New York: McGraw-Hill Book Company.

Connaughton, S.L., and M. Shuffler. 2007. "Multinational and Multicultural Distributed Teams: A Review and Future Agenda." *Small Group Research* 38, no. 3, pp. 387–412. doi: http://dx.doi.org/10.1177/1046496407301970

"Country, Nation or State?" 2011. http://www.catataxis.com/index.php/country-nation-or-state (accessed November 7, 2013).

Crossman, A. 2014. *Convergence Theory*. http://sociology.about.com/od/C_Index/g/Convergence-Theory.htm (accessed January 17).

CSU. 2014. http://hsc.csu.edu.au/hospitality/hosp_240/comp_units/SITXCOM 002A/4114/frontpage.htm (accessed January 25, 2014).

Cummings, J., and C. Pletcher. 2011. "Why Project Networks Beat Project Teams." *MIT Sloan Management Review* 52, no. 3, pp. 75–80.

de Bony, J. 2009. "Project Management and National Culture: A Dutch-French Case Study." *International Journal of Project Management* 28, no. 2, pp. 173–82. doi: http://dx.doi.org/10.1016/j.ijproman.2009.09.002

de Dreu, C.K.W., and L.R. Weingart. 2003. "Task versus Relationship Conflict, Team Performance, and Team Member Satisfaction: A Meta-Analysis." *Journal of Applied Psychology* 88, no. 4, pp. 741–49. doi: http://dx.doi.org/10.1037/0021-9010.88.4.741

de Meyer, A., C. H. Loch, and M.T. Pich. Winter 2002. "Managing Project Uncertainty: From Variation to Chaos." *MIT Sloan Management Review*, 43, no. 2, pp. 60–67.

Deal, T.E., and A.A. Kennedy. 1982. *Corporate Cultures: The Rites and Rituals of Corporate Life*. Boston, MA: Addison-Wesly Publishing Company Inc.

DifferenceBetween.net. 2013. http://www.differencebetween.net/language/difference-between-values-and-beliefs (accessed August 29).

Dunn, S.C. 2001. "Motivation by Project Functional Managers in Matrix Organizations." *Engineering Management Journal* 13, no. 2, pp. 3–9.

Earley, P.C., and S. Ang. 2003. *Cultural Intelligence: Individual Interactions across Cultures*. Redwood City, CA: Stanford University Press.

Earley, P.C., and M. Erez. 1997. *The Transplanted Executive*. New York: Oxford Press.

Elrod, C., A. Rasnic, and W. Daughton. 2007. "Engineering Management and Industrial Engineering: Similarities and Differences." *American Society for Engineering Education*.

Eskerod, P., and H.J. Skriver. 2007. "Organizational Culture Restraining IN-House Knowledge Transfer Between Project Managers - A Case Study." *Project Management Journal*, 38, no. 1, pp. 110–122.

Faulcombridge, R.I., and M.J. Ryan. 2002. *Managing Complex Technical Projects: A Systems Engineering Approach*. Boston, MA: Artech House.

Fellows, R., A. Liu, and C. Storey. 2004. "Ethics in Construction Project Briefing." *Science and Engineering Ethics* 10, no. 2, pp. 289–301. doi: http://dx.doi.org/10.1007/s11948-004-0025-5

Fells, M.J. 2000. "Fayol Stands the Test of Time." *Journal of Management History* 6, no. 8, pp. 345. doi: http://dx.doi.org/10.1108/13552520010359379

Gannon, M.J. 1994. *Understanding Global Cultures: Metaphorical Journey through 17 Countries*. Thousand Oaks, CA: Sage.

GAPPS (Global Alliance for Project Performance Standards). July 2006. *A Framework for Performance Based Competency Standards for Global Level 1 and 2 Project Managers*. GAPPS.

Garvey, M. 2013. http://www.brainyquote.com/quotes/quotes/m/marcus-garv365148.html (accessed August 21).

Gerhart, B. 2008. "How Much Does National Culture Constrain Organizational Culture?" *Management and Organization Review* 5, no. 2, pp. 241–59. doi: http://dx.doi.org/10.1111/j.1740-8784.2008.00117.x

Goethals, G.W., and J.W.M. Whiting. December 1957. "Research Methods: The Cross-Cultural Method." *Review of Educational Research* 27, no. 5, pp. 441–8. doi: http://dx.doi.org/10.3102/00346543027005441

Gray, C.F., and E.W. Larson. 2008. *Project Management: The Managerial Process.* 4th ed. Boston, MA: McGraw-Hill.

Gress, D. 1999. "Multiculturalism in World History." http://www.fpri.org/articles/1999/09/multiculturalism-world-history (accessed January 17, 2014).

Guignery, V., C. Pesso-Miquel, and F. Specq. 2011. *Hybridity: Forms and Figures in Literature and the Visual Arts.* Newcastle, DE: Cambridge Scholars Publishing.

Gunding, E. 2013. *Working GlobeSmart: 12 People Skills for Doing Business across Borders.* Mountain View, CA: Davies-Black Publishing.

Hampden-Turner, C., and F. Trompenaars. 1997. "Response to Geert Hofstede." *International Journal of Intercultural Relations* 21, no. 1, pp. 149–59. doi: http://dx.doi.org/10.1016/s0147-1767(96)00042-9

Handy, C. 1989. "The Future of Project Teams." http://www.thomsett.com/Resources/PDF/The%20future%20of%20project%20teams.pdf (accessed November 9, 2013).

Hare, D., Sir. 2013. "12Manage The Executive Fast Track: Project Management Quotes." http://www.12manage.com/quotes_pp.html (accessed June 10).

Hartman, F., and C. Guss. 1996. "Virtual Teams – Constrained by Technology or Culture?" *Managing Virtual Enterprises: A Convergence of Communications, Computing, and Energy Technologies: IEMC 96 Proceedings, International Conference on Engineering and Technology Management, pp.* 185–90.

Haas, H., and S. Nüesch. 2013. "Are Multinational Teams More Successful?" Leading House Working Paper No. 88. The Swiss Leading House on Economics of Education, Firm Behavior and Training Policies. www.economics-of-education.ch

Heath, C., and D. Heath. 2010. *Switch: How to Change Things When Change is Hard.* New York: Broadway Books.

Heise, D.R. 2013. "Modeling Interactions in Small Groups." *Social Psychology Quarterly* 76, no. 1, pp. 52–72. doi: http://dx.doi.org/10.1177/0190272512467654

Heizer, J., and B. Render. 2006. *Operations Management.* 8th ed. Upper Saddle River, NJ: Pearson Prentice Hall.

Helgadóttir, H. 2008. "The Ethical Dimension of Project Management." *International Journal of Project Management.* 26, no. 7, pp. 743–48. doi: http://dx.doi.org/10.1016/j.ijproman.2007.11.002

Henrie, M. *Multi-National Project Team Communications and Cultural Influences* [Unpublished Dissertation]. Old Dominion University; 2005.

Henrie, M. 2010. *Multi-National Project Team Communications: International Cultural Influences.* Germany: VDM Verlag Dr. Müller.

Henrie, M., and A. Sousa-Poza. 2005. "Project Management: A Cultural Literary Review." *Project Management Journal* 36, no. 2, pp. 5–14.

Hersey, P., and K. Blanchard. 1982. *Management of Organizational Behavior: Utilizing Human Resources.* 4th ed. Englewood Cliffs, NJ: Prentice-Hall Inc.

Ho, C.M.-F. 2011. "Ethics Management for the Construction Industry." *Engineering, Construction and Architectural Management* 18, no. 5, pp. 516–37. doi: http://dx.doi.org/10.1108/09699981111165194

Hofstede, G. 1997. *Cultures and Organizations: Software of the Mind.* New York: McGraw-Hill.

Hofstede, G. 2013. "Research and VSM." http://www.geerthofstede.com/research--vsm (accessed August 3, 2013).

Hoole, C., and Y. du Plessis. July 14–17 2002. "The Development of a Project Management Culture Assessment Framework." Paper Presented at the PMI Research Conference 2002, Seattle, WA. Project Management Institute.

Hopkinson, M. 2010. *The Project Risk maturity Model: Measuring and Improving Risk Management Capability.* Surrey, UK: Ashgate - Gower.

House, R.J., P.J. Hanges, M. Javidan, P.W. Dorfman, and V. Gupta, eds. 2004. *Culture, Leadership, and Organizations: The Globe Study of 62 Societies.* Thousand Oaks, CA: Sage.

Howard-Grenville, J.A. 2006. "Inside the "Black Box": How Organizational Culture and Subcultures Inform Interpretations and Actions on Environmental Issues." *Organization & Environment* 19, no. 1, pp. 46–73. doi: http://dx.doi.org/10.1177/1086026605285739

IJPM (International Journal of Project Management). 2013. http://www.journals.elsevier.com/international-journal-of-project-management (accessed October 8).

IPMA. 2005. "Welcome to the History of IMPA and Its National Associations." *International Project Management Association.* http://ipma.ch/assets/IPMA-40-0-Introduction.pdf (accessed August 3, 2013).

ISO 21500. September, 2012. *Guidance on Project Management.* 1st ed. International Organization of Standards.

Istvan, R.L. 1992. "A New Productivity Paradigm for Competitive Advantage." *Strategic Management Journal* 13, no. 7, pp. 525–37. doi: http://dx.doi.org/10.1002/smj.4250130705

Jacob, N. 2005. "Cross-cultural Investigations: Emerging Concepts." *Journal of Organizational Change Management* 18, no. 5, pp. 514–28. doi: http://dx.doi.org/10.1108/09534810510614986

Jarvenpaa, S.L., and D.E. Leidner. November–December 1999. "Communication and Trust in Global Virtual Teams." *Organization Science* 10, no. 6, pp. 791–815.

Johns, G. April 2006. "The Essential Impact of Context on Organizational Behavior." *Academy of Management* 31, no. 2, pp. 386–408. doi: http://dx.doi.org/10.5465/amr.2006.20208687

Jogulu, U.D, and G.J. Wood. 2008. "At the Heart of Cross-cultural Research: Challenges in Methodological Design." Presented at the Diversity08, Eighth International Conference on Diversity in Organisations, Communities & Nations. Montreal, Canada, June 17–20.

Karlsen, J.T. 2011. "Supportive Culture for Efficient Project Uncertainty Management." *International Journal of Managing Projects in Business* 4, no. 2, pp. 240–56. doi: http://dx.doi.org/10.1108/17538371111120225

Katz, N., D. Lazer, H. Arrow, and N. Contractor. 2004. "Network Theory and Small Groups." *Small Group Research* 35, no. 3, pp. 307–32. doi: http://dx.doi. org/10.1177/1046496404264941

Katzenbach, J.R., and D.K. Smith. 1993. *The Wisdom of Teams: Creating the High-performance Organization.* Boston, MA: Harvard Business School.

Kendra, K., and L.J. Taplin. 2004. "Project Success: A Cultural Framework." *Project Management Journal* 35, no. 1, pp. 30–45.

Killen, C.P., and C. Kjaer. 2012. "Understanding Project Interdependencies: The Role of Visual Representation, Culture and Process." *International Journal of Project Management* 30, no. 5, pp. 554–66. doi: http://dx.doi. org/10.1016/j.ijproman.2012.01.018

King, S.K. 1997. "Rituals and Modern Society." http://www.huna.org/html/ skritual.html (accessed September 6, 2013).

Kloppenborg, T.J., and W.A. Opfer. 2002. "Forty Years of Project Management Research Trends." In *Interpretations, and Predictions. The Frontiers of Project Management Research*, eds. D.P. Slevin, D.L. Cleland, and J.K. Pinto. Newton Square, PA: Project Management Institute.

Kraidy, M.M. 2002. "Hybridity in Cultural Globalization." *Communication Theory* 12, no. 3, pp. 316–39. doi: http://dx.doi.org/10.1093/ct/12.3.316

Kuhlmann, A. March–April 2010. "Culture-driven Leadership." *Ivey Business Journal Online*, http://search.proquest.com.proxy.consortiumlibrary.org/doc view/347868156 (accessed August 2013).

Lane, G. 2013. "Culturally Aligned." *Industrial Engineer* 45, no. 8, pp. 49–53.

"Latin Word Study Tool." 2013. http://www.perseus.tufts.edu/hopper/morph?l=- natus&la=la (accessed November 7).

Lehman, A. 2010. "Human Evolution: Evolution and the Structure of Health and Disease." http://serpentfd.org/humanevolutionintro.html (accessed August 22, 2013).

Letiecq, B. August 2004. "Evaluating From the Outside Conducting Cross-cultural Evaluation Research on an American Indian Reservation." *Evaluation Review* 28, no. 4, pp. 342–57. doi: http://dx.doi.org/10.1177/0193841x04265185

Levardy, V., and T.R. Browning. November 2009. "An Adaptive Process Model to Support Product Development Project Management." *IEEE Transactions of Engineering Management* 56, no. 4, pp. 600–20. doi: http://dx.doi.org/10.1109/ tem.2009.2033144

Levine, J.M., and R.L. Moreland. 1990. "Progress in Small Group Research." *Annual Review of Psychology* 41, no. 1, pp. 585–634. doi: http://dx.doi. org/10.1146/annurev.psych.41.1.585

Liamputtong, P. 2013. *Doing Research in a Cross-cultural Context: Methodological and Ethical Challenges.* Netherlands: Springer.

Lim, E.C., and J. Alum. 1995. "Construction Productivity: Issues Encountered by Contractors in Singapore." *International Journal of Project Management* 13, no. 1, pp. 51–58. doi: http://dx.doi.org/10.1016/0263-7863(95)95704-h

Lim, L., and P. Firkola. 2000. "Methodological Issues in Cross-cultural Management Research: Problems, Solutions, and Proposals." *Asia Pacific Journal of Management* 17, no. 1, pp. 133–54. doi: http://dx.doi.org/10.1023/A:1015493005484

Locke, E.A. 1982. "The Ideas of Frederick W. Taylor: An Evaluation." *Academy of Management Review* 7, no. 1, pp. 14–24. doi: http://dx.doi.org/10.2307/257244

Loosemore, M., and P. Lee. 2002. "Communication Problems with Ethnic Minorities in the Construction Industry." *Project International Journal of Project Management* 20, no. 7, pp. 517–24. doi: http://dx.doi.org/10.1016/s0263-7863(01)00055-2

Lu, L.-T. 2012. "Etic or Emic? Measuring Culture in International Business Research." *International Business Research* 5, no. 5, pp. 109–15. doi: http://dx.doi.org/10.5539/ibr.v5n5p109

Machiavelli. 2013. "12 Manage the Executive Fast Track; Project Management Quotes." http://www.12manage.com/quotes_pp.html (accessed October 6).

Mahoney, J., and G. Goertz. 2006. "A Tale of Two Cultures: Contrasting Quantitative and Qualitative Research." *Political Analysis* 14, no. 3, pp. 227–29. doi: http://dx.doi.org/10.1093/pan/mpj017

Margolis, S.L. 2013. "What is the Philosophy of an Organization?" http://www.sheilamargolis.com/organizational-culture/the-five-ps-and-organizational-alignment/philosophy (accessed August 29, 2013).

Marrone, J.A. 2010. "Team Boundary Spanning: A Multilevel Review of Past Research and Proposals for the Future." *Journal of Management* 36, no. 4, pp. 911–40. doi: http://dx.doi.org/10.1177/0149206309353945

Maylor, H. February 2001. "Beyond the Gantt Chart: Project Management Moving on." *European Management Journal* 19, no. 1, pp. 92–100. doi: http://dx.doi.org/10.1016/s0263-2373(00)00074-8

McGrath, J.E., and D.H. Gruenfeld. 1993. "Toward a Dynamic and Systemic Theory of Groups: An Integration of Six Temporally Enriched Perspectives." In *The Future of Leadership Research: Promise and Perspective*, 2–46. Orlando, FL: Academic Press.

Merriam-Webster. 2013. http://www.merriam-webster.com/dictionary/ethic (accessed November 28).

Merriam-Webster. 2014. http://www.merriam-webster.com/dictionary/multinational (accessed January 11).

Minkov, M., and G. Hofstede. 2011. "The Evolution of Hofstede's Doctrine." *Cross Cultural Management* 18, no. 1, pp. 10–20.

Minniti, M. 2006. "Entrepreneurs Examined." *Business Strategy Review* 17, no. 4, pp. 78–82.

Molander, E.A. 1987. "A Paradigm for Design, Promulgation, and Enforcement of Ethical Codes." *Journal of Business Ethics* 6, no. 8, pp. 619–31. doi: http://dx.doi.org/10.1007/bf00705778

Monhor, D. December 2011. "A New Probabilistic Approach to the Path Criticality Ion Stochastic PERT." *Central European Journal of Operations Research* 19, no. 4, pp. 615–33. doi: http://dx.doi.org/10.1007/s10100-010-0151-x

Morgan, B.B, Jr., A.S. Glickman, A.E. Woodward, A. Blaiwes, and E. Salas. 1986. Measurement of Team Behaviors in a Navy Environment. NTSC Report No. 86-014. Orlando, FL: Naval Training System Center.

Morrill, C. September, 2008. "Culture and Organization Theory." *Annals of the American Academy of Political and Social Science* 619, no. 1, pp. 15–40. doi: http://dx.doi.org/10.1177/0002716208320241

Morris, M.W., K. Leung, D. Ames, and B. Lickel. 1999. "Views from Inside and Outside: Integrating Emic and Etic Insights about Culture and Justice Judgment." *Academy of Management Review* 24, no. 4, pp. 781–96. doi: http://dx.doi.org/10.5465/amr.1999.2553253

Morton, D.H. 1975. "Project Manager, Catalyst to Constant Change: A Behavioral Analysis." *Project Management Quarterly* 6, no. 1, pp. 22–33.

Müller, R., E.S. Andersen, Ø. Kvalnes, J. Shao, S. Sankaran, J.R. Turner, C. Biesenthal, D. Walke, and S. Gudergan. 2013. "The Interrelationship of Governance, Trust and Ethics in Temporary organizations." *Project Management Journal*. 44, no. 3, pp. 26–44. doi: http://dx.doi.org/10.1002/pmj.21350

Müller, S.D., P. Kraemmergaaard, and L. Mathiassen. 2009. "Managing Cultural Variation in Software Process Improvement: A Comparison of Methods for Subculture Assessment." *IEEE Transactions of Engineering Management* 56, no. 4, pp. 584–99. doi: http://dx.doi.org/10.1109/tem.2009.2013829

Muller, R., and J.R. Turner. 2004. "Cultural Differences in Project Owner-Project Manager Communications." In *Innovations: Project Management Research 2004*, eds. D.P. Slevin, D.L. Cleland, and J.K. Pinto. Newton Square, PA: Project Management Institute.

Munk-Madsen, A. n.d. "The Concept of 'Project': A Proposal for a Unifying Definition."

Nicholas, J.M. 2001. *Project Management for Business and Technology: Principles and Practice.* 2nd ed. Upper Saddle River, NJ: Prentice-Hall Inc.

Nkwi, P., I. Nyamongo, and G.W. Ryan. 2001. *Field Research into Socio-cultural Issues: Methodological Guidelines."* Yaoundé, Cameroon, Africa: International Center for applied Social Sciences, Research, and Training/UFPA.

O'Neil, D. 2006. "What is Culture?" http://anthro.palomar.edu/culture/culture_1.htm (accessed August 22, 2013).

Ochieng, E.G., and A.D.F. Price. 2010. "Managing Cross-cultural Communication in Multicultural Construction Project Teams: The Case of Kenya and UK." *International Journal of Project Management* 28, no. 5, pp. 449–60. doi: http://dx.doi.org/10.1016/j.ijproman.2009.08.001

Online Etymology Dictionary. 2013. http://www.etymonline.com/index.php?term=culture (accessed August 3).

"Our Publications." 2013. PMI. http://www.pmi.org/Resources/Pages/Our-Publications.aspx (accessed October 8, 2013).

Oxford Dictionary. 2013. http://oxforddictionaries.com/us/definition/american_english/project (accessed June 10).

Paglia, C. 2013. http://famous-quotes.com/author.php?aid=5528 (accessed October 25).

Park, H., V. Ribiere, and W.D. Schulte, Jr. 2004. "Critical Attributes of Organizational Culture that Promote Knowledge Management Technology Implementation Success." *Journal of Knowledge Management* 8, no. 3, pp. 106–17. doi: http://dx.doi.org/10.1108/13673270410541079

Pence, K.R., and C.J. Rowe. May–June 2012. "Enhancing Engineering Education through Engineering Management." *Journal of STEM Education: Innovations and Research* 13, no. 3, pp. 46–51.

Peters, T.J. 1979. "Beyond the Matrix Organization." *Business Horizons* 22, no. 5, pp. 15–27. doi: http://dx.doi.org/10.1016/0007-6813(79)90027-2

Peterson, R. Summer 1965. "Critical Path Scheduling: A Comprehensive Look." *Business Quarterly*, pp. 70–86.

Pinto, J.K., and D.P. Slevin. 1989. "Critical Success Factors in R&D Projects." *Research Technology Management* 32, no. 1, pp. 31–34.

PM Network. 2013. http://www.pmi.org/Knowledge-Center/Publications-PM-Network.aspx (accessed October 10).

PMI (Project Management Institute). 1994. *A Guide to the Project Management Body of Knowledge (PMBOK) Exposure Draft.* Newton Square, PA: Project Management Institute.

PMI (Project Management Institute). 1996. *A Guide to the Project Management Body of Knowledge.* Newton Square, PA: PMI.

PMI (Project Management Institute). 2004. *A Guide to the Project Management Body of Knowledge.* Newton Square, PA: PMI.

PMI (Project Management Institute). 2008. *A Guide to the Project Management Body of Knowledge.* Newton Square, PA: PMI.

PMI (Project Management Institute). 2013a. *Project Management between 2010 – 2020.* Newtown Square, PA: PMI.

PMI (Project Management Institute). 2013b. *Project Management Body of Knowledge.* 5th ed. Newtown Square, PA: PMI.

PMIJ. 2014. http://www.pmi.org/Learning/publications-project-management-journal.aspx (accessed October 10).

Ptagorsky, G. 1998. "The Project Manager/Functional Manager Partnership." *Project Management Journal* 29, no. 4, pp. 7–17.

Ralston, D.A. 2008. "The Crossvergence Perspective: Reflections and Projections." *Journal of International Business Studies* 39, no. 1, pp. 27–40. doi: http://dx.doi.org/10.1057/palgrave.jibs.8400333

Rasmussen, P.R. 2001. "'Nations' or 'States' an Attempt at Definition." http://www.globalpolicy.org/component/content/article/172/30341.html (accessed November 7, 2013).

Reid, R.D., and N.R. Sanders. 2005. *Operations Management: An Integrated Approach*. 2nd ed. Canvers, CA: John Wiley & Sons Inc.

Rich, P. 1992. "The Organizational Taxonomy: Definition and Design." *Academy of Management Review* 17, no. 4. pp. 758–81. doi: http://dx.doi.org/10.5465/amr.1992.4279068

Rodwell, J.J., R. Kienzle, and M.A. Shadur. 1998. "The Relationships among Work-Related Perceptions, Employee Attitudes, and Employee Performance: The Integral Role of Communication." *Human Resource Management* 37, no. 34, pp. 277–93. doi: http://dx.doi.org/10.1002/(sici)1099-050x(199823/24)37:3/4<277::aid-hrm9>3.3.co;2-5

Rokeach, M. 1973. *The Nature of Human Values*. New York: The Free Press.

Rubinstein, N. 2004. "Studies in Italian History in the Middle Ages and the Renaissance." In *Political Thought and the Language of Politics*, Vol. 1. Rome, Italy: Storia e Letteratura.

Russell, R.S., and B.W. Taylor III. 2009. *Operations Management: Creating Value along the Supply Chain*. 6th ed. Danvers, MA: John Wiley & Sons Inc.

Ruth, B. 2013. http://www.brainyquote.com/quotes/keywords/team.html (accessed December 26).

Sackmann, S.A. 1992. "Culture and Subcultures: An Analysis of Organizational Knowledge." *Administrative Science Quarterly* 37, no. 1, pp. 140–61. doi: http://dx.doi.org/10.2307/2393536

Schein, E.H. 1990. "Organizational Culture." *American Psychologist* 45, no. 2, pp. 109–10.

Schein, E.H. 2004. *Organizational Culture and Leadership*. 3rd ed. Hoboken, NJ: John Wiley & Sons Inc.

Schermerhorn, J.R., and M.H. Bond. 1997. "Cross-cultural Leadership Dynamics in Collectivism and High Power Distance Settings." *Leadership & Organization Development Journal* 18, no. 4, pp. 187–93. doi: http://dx.doi.org/10.1108/01437739710182287

Shah, H. 2012. *A Guide to the Engineering Management Body of Knowledge*. 3rd ed. Rolla, MO: The American Society for Engineering Management.

Shore, B. 2008. "Systematic Biases and Culture in Project Failures." *Project Management Journal* 39, no. 4, pp. 5–16. doi: http://dx.doi.org/10.1002/pmj.20082

Shore, B., and B.J. Cross. 2004. "Exploring the Role of National Culture in the Management of Large-scale International Science Projects." *International Journal of Project Management* 23, no. 1, pp. 55–64. doi: http://dx.doi.org/10.1016/j.ijproman.2004.05.009

Sinnette, J. July–August 2004. "Accounting for Megaproject Dollars." *Public Roads* 68, no. 1.

Smith, S.M. February 2010. "Finding the Right Fit: How to Leverage Culture to Drive Project Team Success." *Business Strategies*, pp. 20–25.

Smits, K. *Cross Culture Work: Practices of Collaboration in the Panama Canal Expansion Program* [Unpublished dissertation]. Delft, Netherlands: Next Generation Infrastructures Foundation; 2013.

Stake, R.E. 1995. *The Art of Case Study Research.* Thousand Oaks, CA: Sage Publications.

Stallsworth, E. May 2009. "Using a PERT Chart." Bright Hub PM. http://www.brighthubpm.com/project-planning/4997-using-a-pert-chart (accessed June 23, 2009).

Starks, R.R., J. McCormack, and S. Cornell. 2013. "Native Nations and U.S. Boarders: Challenges to Indigenous Culture, Citizenship, and Security." Tucson: The University of Arizona. http://nni.arizona.edu/pubs/Native%20Nations%20and%20US%20Borders_sample%20chapter.pdf (accessed November 8, 2013).

Storti, C. 1994. *Cross-Cultural Dialogues: Brief Encounters with Cultural Difference.* Yarmouth, MA: Intercultural Press.

Storti, C. 1998. *Figuring Foreigners Out.* Boston, MA: Intercultural Press Inc.

Suran, S. 2003. "How to Implement Change Effectively." *Journal of Corporate Accounting & Finance* 14, no. 2, pp. 31–38. doi: http://dx.doi.org/10.1002/jcaf.10134

Tamu. 2013. http://www.tamu.edu/faculty/choudhury/culture.html (accessed September 3).

Taylor, E.B. 1871. "Primitive Culture: Researches into the Development of Mythology, Philosophy, Religion, Language, Art, and Custom." London, UK: John Murray. http://archive.org/details/primitiveculture01tylouoft (accessed August 22, 2013).

The American Heritage® Dictionary of the English Language. 2001. 4th ed.

Toledo. 2014. http://psychology.utoledo.edu/showpage.asp?name=SCIRL_home (accessed January 14).

Trompenaars, F., and C. M., Hampden-Turner. 1998. *Riding the Waves of Culture: Understanding Diversity in Global Business.* New York: McGraw-Hill.

Velasquez, M., C. Andre, T. Shanks, and M.J. Meyer. 2014. "What is Ethics?" http://www.scu.edu/ethics/practicing/decision/whatisethics.html (accessed January 24).

Wang, X., and L. Liu. 2007. "Cultural Barriers to the Use of Western Project Management in Chinese Enterprises: Some Empirical Evidence from Yunnan Province." *Project Management Journal* 38, no. 3, pp. 61–73. doi: http://dx.doi.org/10.1002/pmj.20006

Webster, J., and W.K.P. Wong. 2008. "Comparing Traditional and Virtual Group Forms: Identify, Communication and Trust in Naturally Occurring Project Teams." *The International Journal of Human Resource Management* 19, no. 1, pp. 41–62. doi: http://dx.doi.org/10.1080/09585190701763883

Weitz, L. 1993. "Merging Process, Project Management." *Software Magazine* 13, no. 8, pp. 67.

"What Are Dueling Maxims?" 2013. About Education. http://grammar.about.com/od/qaaboutrhetoric/f/maximqa.htm (accessed November 29).

"When Did Globalisation Start?" 2014. *The Economist.* http://www.economist.com/blogs/freeexchange/2013/09/economic-history-1 (accessed January 17, 2014).

Wiewiora, A., B. Trigunarsyah, G. Murphy, and V. Coffey. November 2013. "Organizational Culture and Willingness to Share Knowledge: A Competing Values Perspective in Australian Context." *International Journal of Project Management* 31, no. 8, pp. 1163–74. doi: http://dx.doi.org/10.1016/j.ijproman.2012.12.014

Wikipedia. 2013. http://en.wikipedia.org/wiki/Federation (accessed October 9).

Williams, R. 2013. "Excerpts from Raymond Williams, *Keywords*." http://pubpages.unh.edu/~dml3/880williams.htm (accessed August 22, 2013).

Williams, W. 2013. http://gcimmarrusti.wordpress.com/pm-quotes (accessed November 12).

Wilson, J.M. 2003. "Gantt charts: A Centenary Appreciation." *European Journal of Operational Research* 149, no. 2, pp. 430–7. doi: http://dx.doi.org/10.1016/s0377-2217(02)00769-5

World Bank. 2013. http://web.worldbank.org/WBSITE/EXTERNAL/TOPICS/ENVIRONMENT/EXTENVASS/0,,menuPK:407994~pagePK:149018~piPK:149093~theSitePK:407988,00.html (accessed November 29).

World Forum. 2000. http://www.globalpolicy.org/globalization/defining-globalization/48051.html (accessed January 18, 2014).

Wrege, C.D., and A.M. Stotka. 1978. "Cooke Creates a Classic: The Story Behind F. W. Taylor's Principles of Scientific Management." *The Academy of Management Review* 3, no. 4, pp. 736–49. doi: http://dx.doi.org/10.2307/257929

Xiuyan, X. 2012. "Cultural Factors in EAP Teaching—Influences of Thought Pattern on English Academic Writing." *Cross-Cultural Communication* 8, no. 4, pp. 53–57. doi: http://dx.doi.org/10.3968%2Fj.ccc.1923670020120804.2307

Yan, C. 2009. "Convergence and Divergence: Examine Perceptions of Chinese and Expatriate Project Implementers on Cross-cultural Teacher Training Programmes." *Compare: A Journal of Comparative and International Education* 39, no. 6, pp. 691–706. doi: http://dx.doi.org/10.1080/03057920902834261

Yazici, H.J. June 2011. "Significance of Organizational Culture in Perceived Project and Business Performance." *Engineering Management Journal* 23, no. 2, pp. 20–29.

Zeitun, R.M., K.S. Abdulgader, and K.A. Alshare. 2013. "Team Satisfaction and Student Group Performance: A Cross-cultural Study." *Journal of Education for Business* 88, no. 5, pp. 286–93. doi: http://dx.doi.org/10.1080/08832323.2012.701243

Zheng, W., B. Yang, and G.N. McLean. 2010. "Linking Organizational Culture, Structure, Strategy, and Organizational Effectiveness: Mediating Role of Knowledge.

Zwikael, O., K. Shimizu, and S. Globerson. 2005. "Cultural Differences in Project Management Capabilities: A Field Study." *International Journal of Project Management* 23, no. 6, pp. 454–62. doi: http://dx.doi.org/10.1016/j.ijproman.2005.04.003

BIBLIOGRAPHY

Ilgen, D.R. February 1999. "Teams Embedded in Organizations." *American Psychologist* 54, no. 2, pp. 129–39. doi: http://dx.doi.org/10.1037//0003-066x.54.2.129

Jetu, F.T., R. Ridel, and F. Roithmay. September 2010. "Culture Patterns Influencing Project Team Behavior in Sub-Saharan Africa: A Case Study in Ethiopia." *Project Management Journal* 42, no. 5, pp. 57–77. doi: http://dx.doi.org/10.1002/pmj.20260

Kelley, J.E., Jr. May–June 1961. "Critical-Path Planning and Scheduling: Mathematical Basis." *Operations Research* 9, no.3, pp. 296–320. doi: http://dx.doi.org/10.1287/opre.9.3.296

MA. "Answers: What is IPMA?" *International Project Management Association,* http://ipma.ch/about/answers (accessed August 3, 2013).

Moder, J.J., C.R. Phillips, and E.W. Davis. 1995. *Project Management with CPM, PERT and Precedence Diagramming.* 3rd ed. Middleton, MA: Blitz Publishing Company.

PMIJ. 2013. http://www.pmi.org/Knowledge-Center/Publications-Project-Management-Journal.aspx (accessed October 10).

Saradar, Z., and B. van Loon. 1997. *Cultural Studies for Beginners.* Australia: Icon Books.

Schein, E.H. 1992. *Organizational Culture and Leadership.* San Francisco, CA: Jossey-Bass.

Thamhain, H.J. 2004. "Linkages of Project Environment to Performance: Lessons for Team Leadership." *International Journal of Project Management* 22, no. 7, pp. 533–44. doi: http://dx.doi.org/10.1016/j.ijproman.2004.04.005

Wang, X. 2001. Identification and Evaluation of the Key Attributes of Project Management Culture *[Unpublished doctoral thesis]. Melbourne, Australia: School of Management Victoria University of Technology; 2001.*

About the Author

Dr. Morgan Henrie obtained his PhD in Engineering Management and Systems Engineering from Old Dominion University. His dissertation focused on multinational team communications and cultural influences. He is the author of over a 100 papers, articles, and book chapters. Dr. Henrie lives in Anchorage, Alaska, and travels extensively for work and pleasure.

INDEX

THIS BOOK IS IN OUR INDUSTRIAL, SYSTEMS, AND INNOVATION ENGINEERING COLLECTION

William R. Peterson, Editor

Momentum Press is dedicated to developing collections of complementary titles within specific engineering disciplines and across topics of interest. Each collection is led by a collection editor or editors who actively chart the strategic direction of the collection, assist authors in focusing the work in a concise and applied direction, and help deliver immediately actionable concepts for advanced engineering students for course reading and reference.

Some of our collections include:

- *Manufacturing and Processes*—Wayne Hung, Editor
- *Engineering Management*—Carl Chang, Editor
- *Electrical Power*—Hemchandra M. Shertukde, Editor
- *Communications and Signal Processing*—Orlando Baiocchi, Editor
- *Electronic Circuits and Semiconductor Devices*—Ashok Goel, Editor
- *Thermal Science and Energy Engineering*—Derek Dunn-Rankin, Editor
- *Fluid Mechanics*—George D. Catalano, Editor

- *Environmental Engineering*—Francis Hopcroft, Editor
- *Geotechnical Engineering*—Hiroshan Hettiarachchi, Editor
- *Transportation Engineering*—Bryan Katz, Editor
- *Sustainable Structural Systems*—Mohammad Noori, Editor
- *Chemical Reaction Engineering*
- *Chemical Plant and Process Design*
- *Petroleum Engineering*
- *Materials Characterization and Analysis*—Richard Brundle, Editor
- *Computational Materials Science*

Momentum Press actively seeking collection editors as well as authors. For more information about becoming an MP author or collection editor, please visit http://www.momentumpress.net/contact

---

Announcing Digital Content Crafted by Librarians

Momentum Press offers digital content as authoritative treatments of advanced engineering topics by leaders in their field. Hosted on ebrary, MP provides practitioners, researchers, faculty, and students in engineering, science, and industry with innovative electronic content in sensors and controls engineering, advanced energy engineering, manufacturing, and materials science.

Momentum Press offers library-friendly terms:

- perpetual access for a one-time fee
- no subscriptions or access fees required
- unlimited concurrent usage permitted
- downloadable PDFs provided
- free MARC records included
- free trials

The **Momentum Press** digital library is very affordable, with no obligation to buy in future years.

For more information, please visit **www.momentumpress.net/library** or to set up a trial in the US, please contact **mpsales@globalepress.com**.